Connecting With Sounds:
A Network History

Richard

Table of Contents

Introduction

In the late 1990s, my father installed dial-up internet in our house. It was an exciting event. As soon as I was of an age they deemed appropriate, my parents allowed me to use the internet to do research and school work as necessary. I distinctly remember two aspects of this experience: first, I always had to make sure that no one was on the landline before I connected, and that no one was planning to use the phone or was expecting a call; second, the computer made some noise—a dial tone, some beeping, staticky noises, scratchy sounds, among other weird sounds—and then silence, which was the signal for a successful connection.[1] These were two of the most important bits of information that people connecting to the internet needed to know (or hear) at that time. They did not need to concern themselves with the technical particulars of the process at hand. They did not need to know what each machine-made sound meant, either, even though they knew that no sound meant "success." That is, during the dial-up era, people were tacitly trained to listen for specific sounds—sound events, if you will—in relation to networks. Similarly, people today may expect a phone call or text message on their phones when they hear a staticky buzz on their speakers; the closer the phone is to the speaker, the louder the buzz.

With time, and as networks become more silent and seamless, listening for network events is less of a worry for people. In fact, networks today are designed to run in the background, calmly and magically. If, for any reason, their sounds surface in our acoustic sphere, then something must be wrong. This dissertation invites readers to listen to networks, perhaps

---

[1] A quick search of "dial-up tone" on YouTube results with a number of examples.

again or for the first time, in order to investigate what they may learn about network communications while listening to their sounds and monitoring their rhythms.

The following chapters argue that the importance of such listening lies in its potential to help people tell their own "network stories." *Network stories are stories that people tell about networks, network communications, and technologies that rely on networks to function* (including but not limited to "smart" technologies and telecommunications technologies). Contemporary network stories found in the media usually embrace the invisible, inaudible, and non-tangible aspect of networks, to drive home ideas that induce fear and doubt. While such stories invoke serious concerns around privacy and security, they do so in ways that alienate people from the systems in question, without allowing them to first consider how networks manifest in their daily lives. For example, the recently published Netflix documentary, *The Social Dilemma* (Orlowski 2020), creates a parallel, semi-fictional storyline that culminates with a teenager and his sister getting wrongfully arrested at a rally. In that scenario, the viewer sees three identical adult characters who manipulate the boy through his phone, specifically his social media feeds and messages. This sort of anthropomorphized network reduces the complexity of the processes involved, and highlights the "evil" of such systems by attending to suspect human behaviour. Beyond this imagery, people are left with abstract concepts and jargon communicated by interviewees; those communications are in turn manipulated and amplified by the documentary's producers.

As this dissertation demonstrates, the problem with most network stories in popular media is that they tend to ignore the materiality of networks and use abstraction to further expand the gap between people, infrastructure, and communications. To address this problem, and to investigate the materiality of networks in people's everyday lives, I propose an alternative approach involving the use of sonification and Pauline Oliveros's practice of "deep listening"

(Oliveros 2005). This approach relies on the fact that network communications are established

through electromagnetic (EM) waves that travel between networked devices, carrying

algorithmic messages and commands: "[a] world with mobile phones," for example, "would not

be possible without one resource essential to all wireless communications: EM energy. It is EM

energy that enables us to broadcast radio signals over the air" (Gow and Smith 5). Although

these waves and vibrations are inaudible and invisible to people, transduction devices—such as

the "Detektor" used for this dissertation—make it possible for people to record and hear what the

EM waves sound like when converted into decibel levels audible to the human ear.

Network Stories and Network Communications

In a network, algorithms shape how people and technologies relate; this relationship is

frequently described through terms such as "smart devices" or "smart cities," where the word

"smart" connotes innovation as well as the technology's autonomy. To engage this discourse,

this dissertation addresses network stories about present-day networks, where people tell stories

in ways that disregard and migrate from the complexities inherent to network communications.

The discrepancy between how networks communicate and what people know about their

composition reinforces power and control dynamics between the public and the industry, or

between people who are technology experts and those who are not. These dynamics are

particularly oppressive to those who rely on networks, unaware of the possible implications of

their failures. They are even more oppressive when networks behave exactly how they were

designed to behave. Yet, when told differently—be it conscientiously, critically, or

responsibly—network stories have the capacity to help people better understand power dynamics

and, in some instances, provide accurate and understandable details around matters of privacy and security. In other cases, network stories can alert people to the complexities of networks and how they manifest in everyday life, particularly by taking part in the "power play" of technologies and culture, as discussed in chapter 1 of this dissertation.

Network Stories

Network stories—or stories *about* networks—are identifiable for their use of real-life examples, peppered with some fiction and technical details, of an instance or situation when networks and their algorithms either raise questions for the public or cause tangible damage (be it financial or operational) to the parties involved. A network story is usually communicated through the media (such as the British Broadcasting Corporation [BBC] or Canadian Broadcasting Corporation [CBC]) using rhetorical conventions associated with journalism. These stories tend to be structured around a larger, real-life event, with a sequence of events embedded within it. These smaller events are mostly problems or complications, some of which could even be read as disastrous. Network stories do not necessarily contain solutions. In fact, they may strategically avoid them. Either way, they leave audiences with many questions and doubts. Shintaro Miyazaki, for instance, analyzes the network story of the American Telephone and Telegraph Company (AT&T) network crash on January 15, 1990. A programming mistake led one of the network's computers to shut down. Since the system was networked, this problem was communicated to neighbouring machines, so that they would avoid exchanging data with the shutdown computer. However, every other machine then shut down while updating the routing map, all in an attempt to avoid the one problematic machine as they reconnected to the network. The programming mistake caused the entire network to be down for nine hours and "half of the long distance calls [went] unconnected," leading to a loss of more than $60 million, plus the direct financial damage AT&T endured, as well as countless interrupted communications between people (Miyazaki 133). The disconnections affected not only the company itself, but also clients who relied on AT&T for their businesses and everyday lives. Although the financial damage may take precedence in such a network story, people's lives are interrupted in ways that are not necessarily subject of discussion—at least not in the media.[2]

A more recent story attests to how networks affect everyday life as they become more prominent in urban planning. In partnership with Toronto Waterfront, Sidewalk Labs (a Google sister company) had been planning a smart city across twelve acres on the Toronto waterfront, all before they had to shut down in February 2020. Ken Greenberg, Toronto's Former Director of Urban Design and Architecture, claimed the following: "Sidewalk Labs is using technology as a way of enhancing human interaction [and] enhancing community" (Mecava 2017). This remark aligns with Mark Weiser and John Seely Brown's assertion that calm technology enriches "our opportunities for being with other people" (85). Put differently, smart technologies are designed to work like magic, calmly in the periphery of people's attention, allowing them to focus more on their daily activities and interconnections with one another (Weiser 71). While many designers and developers may approach such a scenario optimistically, the Toronto Waterfront project was ultimately surrounded by scepticism and inundated with ethical questions. The BBC

---

[2] When a telephone network is down, people's personal communications are interrupted. For example, contacting a loved one or trying to reach a friend would not be possible during the time when the network is down. While it might not always be too problematic to have a call fail in going through, there are instances where a phone call is of urgent matter: a parent calling their partner because they are unable to pick up the children from school, or a care taker informing a family member that their elder is in the hospital. The network glitch in such cases might result in stressful and at times catastrophic consequences, especially given the fact that such circumstances are personal to people, and the way in which people manage their daily life can vary in terms of emergency planning. With the extent to which our lives today depend on network communications, network glitches are very likely to have effects that reach our very close circles. Even more, people may find themselves living in cities that have entire infrastructures built around networks, which increases the extent to which people's everyday lives are dependant of the successful maintenance of a network.

(2018) explored various concerns raised by Toronto's residents and political figures, highlighting suspicions about privacy issues. One BBC article argued that the project's details about data collection and storage were vague. Further, the CBC (2017) made a curious claim: "To truly build the city of the future from the ground up necessitates that we demolish the city of the past. And that requires some thoughtful considerations about the trade-offs being made, questions that *conveniently* won't need to be asked if Google gets lucky and wins the bid for Toronto's Waterfront" (Pringle 2017; my emphasis). In other words, the fact that Toronto's waterfront was underdeveloped allowed Google and Sidewalk Labs to proceed with their project via "smart gentrification" without much, if any, consideration of historical context and precedence, since the project did not need to destroy existing infrastructure.

An execution copy of the "Plan Development Agreement between Toronto Waterfront Revitalization Corporation and Sidewalk Labs" was released on July 31, 2018.[3] It is more or less a legal document, which uses language to protect Sidewalk in cases of conflict during the project's execution and throughout its lifespan. However, the section on "Digital Governance Framework Principles" does provide some answers, though mostly abstract, to widespread social concerns about privacy and security in the smart city. One of its principles promises "safe use of data [… via] open protocols and rules," while another pledges "algorithmic transparency to avoid bias or marginalization of members or groups of the population" (48). The project also encourages "open innovation" by "[e]nsur[ing] that protocols, standards and operating agreements do not foster monopolies, barriers to entry or lock-in" (49). Such an outreaching call arguably aims to foster a sense of community with prospective investors and "members" of the city, and thus gives the impression of open and transparent practices. Yet the vague and jargon-

---

[3] The Plan Development Agreement was terminated, effective May 17, 2020.

filled nature of the statements suggests otherwise. The language itself functions as a barrier between the company and potential community members, particularly in its undefined use of terms that are not accompanied by any concrete descriptions or proposed practices.

Following the work of both Miyazaki and Alexander Galloway, the Toronto Waterfront story is an example of what Galloway calls "voluntary regulation" (7), inviting people to willingly take part in a network that regulates their behaviours. Relying heavily on dry technical language coupled with vague references to citizen science, the Toronto Waterfront agreement expects "responsible data use" through a "public interest centered approach" (48), placing the responsibility on users instead of technologies. In that sense, it is a neoliberal project because "all aspects of life [are subject] to the so-called discipline of the market" (Shaviro 2015). Networks therefore contribute to, if not enhance, an ideology of constant productivity, where even in times and spaces dedicated to rest and leisure, people contribute to the market in one way or another. The irony, then, is that people's lives are commodified through the promise of a "responsible use of data." While Sidewalk promises "responsible *use* of data," they keep from addressing the irresponsible *data production* pattern to which the community is party, even if unknowingly. In doing so, the residents of the smart city are constantly engaging the market as productive participants in data production, regardless of how their data are used.

Network Communications

While people tell stories *about* networks, networks also have their own way of communicating. Wireless networks, for example, communicate through electromagnetic waves that devices send to and receive from each other to achieve and maintain connections. When a laptop connects to a Wi-Fi router, the phone and router are constantly exchanging EM waves. These streams of communication and EM energy emerge from the mathematical and computational processes of networks and their algorithms. People cannot generally hear these communications without special devices that transduce EM waves into decibel levels that are audible to—and transduced by—human ears. In other words, transduction devices "make sense" (Miyazaki 244) of network connections by rendering their electromagnetic emissions sonically accessible. In doing so, they register a network's patterns of communication and behaviour. These patterns sound like random noise to most people. Audio recordings of them are crowded with multiple sounds layered on top of each other, and transduction devices produce those sounds from a wide range of EM waves. This means that people can only listen to the acousmatics of network infrastructure; they do not have the option to listen to just one network, or to "mute" particular networks in a system. Most important, people are rarely trained to hear the sounds of network communications. As I explain in chapter 3, these sounds are buzzes, scratches, and squeaks that are not common in people's daily lives. Even if they are, they are not meant to "make sense." Since it is difficult, if not impossible, to separate these sounds into their own channels, they become collectively rhythmic in nature, particularly because they echo the infrastructures of networks and algorithmic processes. The resulting rhythms produce a material flow of events that, in their totality, achieve a communication stream.[4] The methodologies at work in this dissertation render those communication streams more tangible, and even comprehensible, for listeners, and they build upon two particular transduction devices: the Detektor and Elektrosluch.

Miyazaki and Martin Howse created the Detektor to listen to the EM emissions of networks and their devices. Jonáš Gruska's Elektrosluch is another such device designed for the

---

[4] While the streams that I record for the purposes of this dissertation are clipped at a certain length, the network communications do not stop; they are continuous even when inaudible.

same purposes. Similar in their composition, the two vary mostly in their aesthetics, though the Elektrosluch has four different iterations. Using either to listen to the sounds that networks emit provides people with material to begin apprehending network communications streams. Many examples of the various uses of the Elektrosluch are available on SoundCloud. Skiesbleed (a SoundCloud user) published two recordings of the Samsung Galaxy S5: the first clip records use of the phone (via touch screen) and its receipt of text messages, and the second is a recording of the same device running but with no user activity.[5] The difference between the two clips is remarkable. The recording of the active device varies in its rhythms, sound levels, and even moments when nothing is heard. Meanwhile, the recording of the inactive device is rhythmic, with very slight variations.

Even without knowing exactly what the user is doing and when, a person listening to these clips can tell when there has been an event, a short action, or an extended and continuous activity. They can identify moments when network communication is established, being maintained, and interrupted. This exercise is very similar in execution and outcome to the learned practice of understanding important event cues when the connection tones of a dial-up modem are sounding. Furthermore, the variations in sounds and their patterns imply different user interactions with the device. These interactions are mechanics that translate into sound events, which are made possible through, and executed according to, sets of algorithms and their protocols. Making EM waves audible thus allows people to hear patterns and sonic events that correspond with the development and execution of certain network communications. Chapter 3 employs this technique of listening to network transduction to investigate how people might begin to tell stories *with and through* networks, not merely *about* them, by using their own communications as a sensible model (McPherson 2009).

The Problem with Talking about Networks

The primary problem this dissertation addresses, especially in chapters 2 and 3, is that the popular stories people tell *about* networks do not necessarily correspond with what and how networks communicate. Given how networks operate and communicate, people are left with a process that is invisible, inaudible, and intangible outside of nonhuman connections, hence the present-day proliferation of network stories in the media amidst the growing number of networks, "smart" projects, and Internet of Things (IoT) infrastructures. Such stories not only cause discrepancies between communications and how people perceive them; they also deter people from intervening in and experimenting with networks. In the context of this dissertation, such intervention means consent to participate, and a willingness to act, in telling network stories. To tell a network story *with and through* networks is to invest in a given system's communications and to employ, in this case, the audible outcomes. It also implies the process of learning the details of a concept and technology (i.e., a network), understanding its social context, monitoring its communications, assessing and interpreting its language and effects, and integrating these notions into other network stories that people are telling.

To arrive at a coherent technique for such an engagement with networks and stories, there are three distinct problems that need to be tackled: a consent problem, a representational problem, and a practical problem. I frame each of these problems within the context of Science and Technology Studies (STS).

Consent Problem

Network stories rehearse what Matthew Kirschenbaum calls a medial ideology (2007),[6] where people do not typically understand how networks function, let alone how their technical particulars shape social relations. People passively give consent to participate in networks; they become features or characters in network stories through the mere fact that they are "members" of a network. Here again, networks are indeed like magic: they leave people with little, if any, access to how they behave. One may argue that the magic of networks is an integral part of their design,[7] "since," as Carolyn Marvin points out, "connections in scientific explanations were often as mysterious to laymen as if they were in fact magic" (113). Attempts at technical explanation can, perhaps ironically, reinforce the rather magical nature of networks, because the general public—or "laymen"—does not have enough access to knowledge that would allow them to make sense of the explanations or stories at hand. To some extent, magic builds dependency, or at least trust, so that people can pursue other aspects of their daily lives. Not having to worry about the complexities or intricacies of how a given network functions allows people to use it and be part of it—one does not need to be a tech savvy to live in a smart city. Nonetheless, investigating the ostensibly invisible, inaudible, or intangible nature of networks addresses aspects of network design that existing STS research either does not tend to consider or fails to translate into methods accessible to the general public.

Representational Problem

Ocularcentric approaches (Jay 1993) to computation focus on abstract forms instead of materiality, and thus tend to also rely heavily on networks as a type of magic. While, again, it is

---

[6] See also: Jerome McGann's *The Romantic Ideology: A Critical Investigation* (1983).
[7] Arthur Clarke's third law of technologies states that "Any sufficiently Advanced Technology is indistinguishable from Magic" (Clarke 21).

sometimes important for networks to function in seamless, calm, and user-friendly ways, the design and discourse of magic usually assumes invisible and presumably non-invasive modes of connection and operation. Yet, computation is not foreign to the materiality of our world. Networks rely on sets of algorithms that are compiled and run on devices to dictate the rules regulating communications. Since these algorithms appear to be intangible or immaterial, they are reified through visualizations. Against these tendencies, listening to networks and their communications—through the amplification of EM emissions, for example—foregrounds and makes palpable the transmissions occurring around us. Analyzing the audible rhythms of network communications also obeys Henri Lefebvre's notion that rhythm "embodies its own law, its own regularity, which it derives from space – from its own space – and from a relationship between space and time" (206). Just as the rhythms of a heart regulate the rhythms of a human body, the rules of algorithms reinforce their rhythms on a network. These rhythms in turn regulate entire systems of networks as they help to establish a network's space-time relationship with the infrastructure in which it operates.

A network operates within the settings of material space-time; it communicates its operation physically by sending and receiving EM waves: devices within a Wi-Fi network send and receive EM waves as per the algorithms (rules) that define how and when devices are allowed to communicate within a particular network. Listening acknowledges the spatio-temporality of patterns without reducing them to visual representations that may conform to a homogenous notion of static or abstract space without rhythm. Equally important, the hope is that listening to networks informs people of the complexities of network communications and prompts them to experiment with how various networks may interact. Intervening in networks this way can provide people with ideas as to how their individual and collective actions affect the behaviour of network communication streams; it translates the materiality of audible networks into human action. Such an apprehension also allows people to sensibly access the extent to which they contribute to networks in the actions they perform and in what they ask of their devices. From wireless connections between devices to touch-screen interactions with phones, people contribute to the rhythms of networks, and these rhythms can be studied through sound. Jacqueline Wernimont, in reference to the "Vibrant Lives" installations she designed with Jessica Rajko, explains that "[i]n creating haptic and sonic experiences of data . . . [o]ur intervention . . . aims to return to individuals a tactile and memorable experience of the data that [was] harvested silently and invisibly, tracking the most mundane of our daily activities" (n.p.). As such, new possibilities arise once formerly cryptic communications become accessible, especially for people who tell stories about networks and aim to responsibly inform the public about them.

With respect to the representational problem, Batya Friedman and Peter Kahn explore the notion of responsible computation, concluding that computer systems can promote responsible computer use by avoiding two features: reducing "the human user to machine-like status" and designing machines to "impersonate human agency" (231). This dissertation focuses on avoiding the latter, applying Friedman and Kahn's framework to the representational problem of ocularcentrism and defining the concept of responsible networks as follows: *a responsible network does not rely on the rhetoric of "giving" people agency; it instead recognizes people as actors whose contributions to a network are essential to the value of its communications and continued connections.*

Practical Problem

Research in STS is useful for moving beyond humanist approaches oriented toward individual practices and objects. The Detektor and the Elektrosluch, for example, both listen to individual devices, with no attention to network communications or any developed technique for employing what is heard. In other words, people do not have a practical means that allows them to listen to how technologies communicate *with* each other in a network; instead, they are left with the vestiges of personal computing practices that privilege the individual and the discrete. The Elektrosluch sound clips available online provide examples of what personal devices sound like; these recordings are similar to those of heart beats monitored via stethoscopes. Alongside them, people would benefit from listening to communications *between* devices and *among* networks, as they would from conversations between two or more people and among groups.

My response to this practical problem is tactical network sonification (TNS), which focuses specifically on rendering networks (not just devices) audible, to then communicate more responsibly about them. I also provide two brief guides on applying and experimenting with the technique, for people who might be interested in producing network sonifications.[8] Although people design and develop devices, networks—in their many layers— often act without us. The magnitude of networks and their infrastructures also makes it impractical for people to approach them device by device, or computer by computer. Acknowledging this magnitude, TNS is a humble contribution to STS, sound studies, and citizen science movements that invite interventions in technologies and systems.

____

Methodology

This dissertation adopts various approaches from materialist and posthumanist strains of STS—and in part critical technical practice (CTP)—as well as sound studies to develop TNS as a methodology. I start by identifying the differences between network communications and network stories, contrasting the layers of network infrastructures and how networks communicate with how networks are perceived as part of everyday life. Once a distinction

Figure 1: Miyazaki and Howse's assembled Detektor, including headphone/amplifier jack, antennas, switch, and other components

between the two is established, I analyze some stories that people tell about networks, including stories about the AT&T crash, the Toronto Smart City, and the Advanced Research Project Agency Network (ARPANet). I focus on events that are particularly vague in stories and kept inaccessible to people, in addition to the events that potentially conflict with the technical and infrastructural particulars of networks, employing Bruno Latour's notion of a "black box" (1988). While the black box expects user contributions to be limited to input and output, this dissertation investigates the materiality of such processes as they unfold beyond the user. To that end, I adopt Beth Coleman's (2011) notion of "x-reality" as the most prominent space for network infrastructure, which "pervasive media use" imposes upon the everyday environment of contemporary society (188).

To study the technical particulars of network protocols and apply a similar method of event analysis, I implement Miyazaki's notion of "algorhythmics" and Oliveros's approach to "deep listening" to conduct TNS. The sonification component of TNS functions as, to quote Marshall McLuhan, an "anti-environment" (McLuhan and Fiore 68) in its ability to alert people to the material presence of network communications among them, even (or particularly) when such communications are silent and invisible by design. Here, it is important to note that the difference between sonic patterns and visual patterns is that the latter flatten variances that are important in terms of EM waves (such as frequency and wavelength). Using Michel Chion's (1994) "reduced listening" (29), I listen to and describe the resulting sounds and noises as a system of networks, avoiding assumptions or declarations about specific devices. Through an attempt at Chion's "causal listening" (25), I acknowledge networks as layered infrastructures, which is a necessary step towards aligning network stories with network communications via an understanding, as per Lefebvre, of rhythms and corresponding events.

When networks communicate, they exchange commands and signals that could announce a network's availability, establish a connection, or maintain a communication stream, among other functions. These events, when translated into sound through TNS, make for good material to understand how networks communicate variations between and across each other and corresponding devices. When people engage in a TNS exercise, where the subject of study may be a local network, they can listen for particular variations, all the while having the opportunity to interfere with active networks—by turning the Wi-Fi router off or even setting a phone on airplane mode, for instance. Within the time and space allocated to such an exercise, people are thus capable of listening intentionally and deeply for fluctuating volumes, frequencies, and rhythms.

Transformation between media and the senses is key to a study of networks; it allows people to approach computation as well as time and space from technical *and* aesthetic perspectives. My research uses a multimodal approach to emphasize the material aspect of information exchange via wireless communications and to better understand and tell network stories. Listening to networks demonstrates that technologies are not as quiet, ephemeral, or immaterial as people may assume them to be. After all, networks adapt to the internal structures of individual technological devices, as well as overall and global network infrastructures, as they afford communication. In other words, the EM exchange that occurs between technologies as they communicate reflects the network protocols and algorithms that allow them to connect devices that do not necessarily share the same forms, platforms, functions, or even communication standards.

As a listening technique, TNS thus primarily involves three steps: 1) investigating network stories, 2) comparative analysis of textual network stories and complex network communications, and 3) sonic anti-environments of networks. In practicing TNS, people would benefit from evaluating the parallels and discrepancies between network communications and how network stories are communicated. In doing so, they can subsequently contribute—in an informed way—to the network processes of everyday environments, as well as responsible telling of corresponding stories. This heuristic also provides people with the tools needed for their practice, both theoretical (i.e. deep listening) and practical. For example, people can listen to the network in their homes to estimate which areas in their households are most crowded with network communications. Another possibility for this method is listening to networked devices, for the purpose of learning which are noisier than others, which in turn informs the listener of the network's potential interruptions and even unsolicited events—such as sonic events that occur when a person is not using any of their devices. Just as dial-up connection sounds allowed people to know when a connection is being established and when it has succeeded, listening to networks that are designed to be silent can produce the same outcome—if not give access to intricacies originally designed to happen beyond people's awareness.

Findings

This dissertation articulates a heuristic for tactical network sonification—as a technique experimenting with McLuhan's anti-environment concept. TNS expands the foundations of STS, aiming to understand the layers of networks and their infrastructures. It provides a different angle than visualizations available in current scholarship that mostly represent networks as a vague concept where "every map of the Internet looks the same" (Galloway 85). McLuhan introduces the term anti-environments as setups or artifacts that are capable of pointing people in the direction of critical engagement with an environment otherwise abstract or unclear (McLuhan and Fiore 68). By listening to network communications, TNSs function as anti-environments and allow people to learn more about network infrastructure, and how networks are situated within our spatio-temporal environment. Sonifications thus direct people's attention (McLuhan and Fiore 68) to the complexities of networks, as a means of better understanding their intricate connections to our everyday life.

Further, TNS provides a definition of responsible technologies, pertinent to networks, inspired by Friedman and Kahn's notion of responsible computation explained above. It also provides technology designers with a theoretical and practical framework that could help guide their designs towards making networks that acknowledge people as active participants. More importantly, it encourages the use of sonifications in support of network stories, or in parallel to network events, to make them less mysterious and more tangible than what most contemporary network stories provide. For instance, in *The Social Dilemma*, instead of using fiction to represent algorithms with three identical men who run a control panel that pushes content and notification to the teenager's phone, story tellers could actually provide clips of what the algorithms that manipulate that teenage boy sound like. They can also employ aesthetics— overlaying music onto the EM clips for example—to make their sonifications more enjoyable to audiences. This approach allows people a more accurate understanding of the complexities of the underlying network communications, all the while relaying the notion of how pervasively present these networks are in people's everyday life.

Chapter Overviews

Chapter 1: A Literature Review of Relevant Scholarship in Science and Technology Studies

The first chapter presents a literature review of relevant scholarship in the field of STS, in support to the background presented in this introduction. It reflects how critical design, materiality, and tactical media come together to provide a reflective and audience-reliant framework for intervening in techno-culture. This chapter highlights the importance of materiality in studying cultural and political aspects of technologies. It also examines how listening provides a way for people to connect networks to their cultural, political, and sociotechnical surroundings. That is, listening as a method carries the rhythms of networks, which are physiologically comprehensible and intellectually intelligible, especially given the effects of rhythms on social behaviour and development (Lefebvre 2013). The notion of "tactical media" as explored by Rita Raley (2009) is the tie that brings together critical theory, design, and materiality, especially in embracing the role of audiences in receiving stories and proceeding with their critical and inquisitive journeys at their own pace. As tactical media encourage disruptive interventions, they also provide people with the language to discuss objects of inquiry—networks in the context of this dissertation—and pursue knowledge in an informed way. While the literature review lays the theoretical grounds for this dissertation, it does not address the technical aspect of sound recordings and EM transductions.

Chapter 2: Rhythms and Network Stories

In this chapter I detail examples of stories that people tell about networks to demonstrate how networks are communicated to the general public. In doing so, I highlight the problematic nature of network stories, particularly their fear inducing rhythms mainly through the use of ambiguity. Alongside the stories about the AT&T crash and Toronto's Sidewalk Smart City, I examine ARPANet, "the first attempt to build a massive computer network, which soon reached geopolitical scales" (Miyazaki 130), as well as ongoing IoT debates (Noble 2018; Coleman 2011). I close-read and rhetorically analyze these stories, grounding them in STS to identify not only the elements that overlap and conflict with the technical particulars of network infrastructures, but more importantly the rhythms that reinforce the social and cultural standings of network stories. Network stories also avoid engaging in a dialogue around networks: chapter 2 explores how even when various parties are involved in an event, the stories do not invoke conversations between weak and strong links, but rather situate them as separate stories about a single network. Such a storytelling approach is not representative of the rhythms of the networks themselves, which contributes to the misconception of pathological systems as opposed to arrhythmic events. In that sense, pathologizing rhythms only reinforces the power structures that the media claim as they compete to remain dominant actors in the realm of networks and network stories.

Chapter 3: Listening for Network Events

In "Listening for Network Events," I experiment with reduced and causal listening, to listen for technological and network events. The outcome of these experiments suggests that it is difficult to isolate networks through their rhythms and sounds. More specifically, listening to network communications entails listening to a system of networks, as opposed to attempting to understand the technical particulars of each network or its components separately. Reduced listening focuses on the sounds one hears outside of their potential causes (Chion 1994). On the other hand, causal listening attempts to identify the cause behind sonic events (Chion 1994). I begin the process by conducting a series of listening exercises, applying reduced and causal listening to readily available transductions of technological events. Then, using Miyazaki and Howse's Detektor, I record five sessions at the University of Victoria's Humanities Computing and Media Centre, transducing EM waves of the network infrastructure of this environment. Using Oliveros's deep listening method (2005), I investigate sonic events and rhythms, to understand network communications. Subsequently, I investigate the role of human and nonhuman communications in the events we hear; as Miyazaki puts it, "[a]lgorithmic ecosystems . . . consist of humans and non-humans, respectively non-machines and machines, constantly exchanging signals not only with one other, but also with their environments and their objects, processes, materials, and bodies" (128). The sound files produced by this project form a corpus that represents algorithmic ecosystems whose presence is invisible. The listening approach is essential because it makes it possible to hear the specific events within networks along with their patterns, so as to make the observations necessary before working towards informed network stories. The third chapter concludes by arguing that to understand networks, it is essential to listen to them as systems, without focusing on the individual networks within an infrastructure.

That said, the algorhythmics of networks make sensible and accessible network events, which are complex in their interconnectedness and intricate in their interdependencies. These events become a representation of the rules that networks impose onto our environments; they are the sounds of the power and control dynamic that is prominent amongst networks, as it gets materially established via EM waves in our everyday life spaces.

Chapter 4: Anti-Environments towards Multimodal Network Stories

In the fourth chapter, I explore the need for listening to network algorhythmics as a technique that allows access to the structures of power and control within networks. As Galloway (2004) explains, algorithms and protocols regulate the operations of network systems; these by extension regulate the EM wave traffic that physically and materially makes technological connections possible. That said, the networks control what Adrian Mackenzie calls the "air interface" (Mackenzie 2010), through which they establish and sustain themselves. Given the material aspect of networks, the architecture of the air interface, and the existence of people within the same environments as networks, people become yet another object that EM waves interact with in establishing network connections. The role of people in networks goes beyond the algorithmic design and making processes. In fact, networks implicate people's everyday life in various operations, not only through their sharing of the air interface and environment, but also through the commands and operations that people run *with* and *through* technological devices within a given network. With that in mind, it becomes vital that people have access to the infrastructural nature of networks to understand how their uses may become an intervention in a given system. I argue that anti-environments of networks act as a diagnostics technique, allowing people to run informed sonic experiments with their everyday network environments. Tactical networks embrace algorhythmics as a method for their design and making processes, to present their findings to the participants in ways that are approachable, and more importantly sensible.

Conclusion

This dissertation concludes by proposing next steps for TNS to reach the general public, particularly because the main concern here is making networks accessible to people outside of academic and industry circles. I use SoundCloud as the main platform on which I publish TNS experiments. There are currently three TNS examples, one of which is a simple overlay of natural ambient sounds on ambient EM waves sound. The other is a piece that intertwines music and the transduction of ambient network sounds at a beach with low EM traffic, and the third presents a similar approach to the second, though in a more technologically crowded space of a public park. These examples aim to introduce people to network communications in provocative ways that raise their curiosity without overwhelming them or inducing fear. Further, examples of TNS function as suggestions for network storytellers, to use such methods in support of their arguments; in particular, when someone is reporting on the failure of a given network, they may choose to contrast a TNS of a working network with one of a failing or interrupted network, for people to hear the complexities that opposing events entail. The conclusion also acknowledges the limitations of this project. This project does not, for instance, address accessibility issues in that it assumes a listening audience. A future goal for TNS is an approach that relies on vibrations via tactile senses—in parallel to sounds—to be more accessible to and inclusive of audiences, such as the Deaf community, who would benefit from such a feature. Lastly, this dissertation encourages its audience to participate in TNS as an anti-environment, to allow for a public contribution to and participation in the networks of everyday life, otherwise restricted to and controlled by the technology industry.

Network communications have become an important and disconcerting part of contemporary discourse. With the exponential reach of social media posts, and the various documentaries and films addressing issues around networks and algorithms, people today are very sceptical of networks, rightfully so. TNS encourages them to listen to networks, as a step towards apprehending how these systems function and communicate. The need for work towards just and equitable systems is very pressing, and this dissertation reinforces this urgency by making sensible the material pervasiveness of networks in our environments. This project is then an invitation for those who might be interested in networks and acousmatics, to take a listen and get a sense of what our technological actions, or the lack thereof, sound like amidst a very crowded infrastructure, especially during a shift imposed on us by a global pandemic. This project also nods at the potential use of aesthetics in communicating complex issues such as networks, by proposing that TNSs have a musical aspect that would be rather essential in making these sonifications approachable to if not, possibly, enjoyable by general audiences.

Chapter 1: A Literature Review of Relevant Scholarship in Science and Technology Studies

Scholarship in science and technology studies has explored the cultural effects of algorithms, and the Internet of Things (IoT), particularly from the perspective of personal computing. This scholarship focuses on code and human-computer interaction (HCI), with less attention to infrastructures and networks. Given the direct effect of personal technologies on people's everyday lives, and the lack of perceived correlation between network communications and people's routines, STS scholarship tends to address the former as it deems them most urgent. Philip Agre's seminal definition of critical technical practice (CTP) highlights the importance of reflecting on technologies—both ideas and practices—as they become increasingly prominent in everyday life (3). Following Agre's investment in artificial intelligence (AI), CTP has proliferated across several fields, with scholars adopting and repurposing it to study an array of technologies and settings (Penny 2000; Ratto 2011; Sengers et al. 2005; Sengers 2005). More specifically, inspired by Agre, critical code studies (CCS), as developed by Mark Marino, sheds light on the non-critical use of personal technologies and the risks it propagates, addressing the issue of people employing what-you-see-is-what-you-get (WYSIWYG) interfaces without much, if any, consideration of the cultural assumptions and values at play (Vee 449).

Marino claims that CCS is interested in the "reception and iteration" of code as a social and material force (473). In doing so, coding as a practice can become a statement as to how technological components affect and are affected by social dimensions. For example, Marino uses the Arabic programming language, *Alb* (Arabic for "heart"), as a case study, and demonstrates how programming languages are written predominantly in English, and how difficult it is to code outside this globalized, colonial framework (475). In other words, code carries language and what cultural ideologies come with it. Similarly, Tarleton Gillespie shows how the main function of algorithms is to calculate values inherent to the designs of algorithms and to provide steps for their own execution (20). With the exponential scale of technology use, these values become disseminated as part of the technology itself. In *Virtual Migration*, A. Aneesh introduces the concept of "algocracy," explaining that code and algorithms shape the construction and persistence of organizational spaces (5). The infrastructure of algorithms is thus by nature one of governance and conditioning. That is, the development of technology via "a disembodied flow of signs and symbols" (Aneesh 9) is not limited to immaterial human labour; the politics of code as an artifact dictate human and machine behaviours all the same (Aneesh 13).

STS publications further explore how technologies regulate societies, through algorithms and technological protocols. These infrastructures are primarily designed to create ordered and controlled environments, within which technologies can coexist. People are added to this equation at a later stage. In *Protocol: How Control Exists after Decentralization,* Alexander Galloway explains that "protocol is a distributed management system that allows control to exist within a heterogeneous material milieu" (7). Seemingly predictable, but incredibly political, protocols are translated and transferred between and within—mostly technological— environments. Yet, for technologies to communicate, algorithms with "shared protocols" are essential and thus definitive for network operations (Galloway 12). Echoing Galloway's early work, Shintaro Miyazaki develops an approach called "algorhythmics." He claims that "'algorhythm' [is] a term that denotes the crucial, dynamic, time critical, real-time manipulative processes that happen when digital technology operates in ultra high-speed operation rates" (141). By combining the words "rhythm" and "algorithm," Miyazaki stresses the temporal aspects of computational processes and suggests we approach algorithms from a structural, social, and material perspective—as opposed to an individual or object-oriented perspective. He elaborates that algorhythms can reveal "breakdowns . . . triggered by algorithms and machinic processes," which cause diverse infrastructural complications (135). Miyazaki conveys examples of such breakdowns and illustrates different aspects of networks and algorithms, the roles they play in everyday life, and how network stories in the media represent these events. Along with other STS research, Miyazaki's work highlights, even when through failure, how technologies become integrated within people's everyday lives, societies, and cultures. More so, STS scholars investigate, through theory and practice, how to critically address and intervene in techno-culture, with and through existing technologies all the while experimenting with the making of new ones. This chapter reviews debates around the following core components: networks, critical technical practice, critical making, algorithm studies, and listening. Debates around these issues set the tone for the discussions in this dissertation, and lay the ground for the methodologies used to develop central claims and arguments. In particular, these conversations provide the framework that I will use to develop a heuristic for listening to networks, as this project's contribution to the STS field.

Defining Networks

Networks are composed of multiple actors and processes, including algorithms, protocols, EM waves, algorithmic events, software, hardware, programmers, and users. Beth Coleman explains that a network

> is a system in which nodes are connected to each other by way of links. In a social network, the nodes would be people. In a biological one, the nodes might be proteins linked to form a metabolic network. Network theory describes a set of behaviors to which a variety of systems—technical, natural, or social—adhere (188).

She then defines networked media as "media platforms that are connected to an informational network, e.g., Internet or cellular towers" (Coleman 188). Conversely, Galloway claims that "[a]ny networked relation will have multiple, nested protocols" (10). As such, a preliminary—though not only—definition of a network is a *system of nodes whose connections are determined and defined by protocols*. Further, Galloway describes the example of the World Wide Web, and in explaining the protocological processes of networks states that

> the entire bundle (the primary data object encapsulated within each successive protocol) is transported according to the rules of the only 'privileged' protocol, that of the physical media itself (fiber-optic cables, telephone lines, air waves). The flexible networks and flows . . . are in fact built directly into the technical specifications of network protocols (11).

This explanation of the process adds a few more layers to the technical complexities of networks: data and physical media. The National Computing Centre defines "*Packet* [as] a block of data handled by a network in a well-defined format including a heading" (82), and a "*Packet Switching Network* [as] a network designed to carry data in the form of packets. The packet and its format is [sic] internal to that network. The external interfaces may handle data in different formats" (82). Packet switching is the communication process within a network, the packets being the data and messages exchanged during a given network communications process. Networks also have hosts—nodes ranging from personal computers to servers to satellites (Galloway 11)—that are also important components of networks. In elaborating on the previous definition, a network, therefore, is a *system of nodes whose connections are determined and defined by protocols, as they exchange packets—of information—in a format determined as per the external interfaces involved.*

Gillespie further explains that an "algorithm . . . is merely the procedure for addressing the task as operationalized: steps for aggregating those assigned values efficiently, or making the matches rapidly, or identifying the stronger relationships according to some operationalized notion of 'strong'" (20). Algorithms thus are the underlying process that determines protocological operations, reinforcing the rules of its "conceptual sequence of steps" (Gillespie 21). More so, "protocol is an algorithm, *a proscription for structure* whose form of appearance may be any number of different diagrams or shapes" (Galloway 30). The algorithmic nature of networks reinforces onto them a structure that is restricted to the rules and regulations of how the network components behave and even manifest. Within these structures, technologies communicate with a degree of flexibility that is ordered by protocols, as they organize and control the flow of communications traffic. Networks that exist and operate within common environments naturally require such an organization. That said, based on existing scholarship, this dissertation proposes the following definition of networks: *networks are an infrastructure for technologies to communicate, determined and prescribed by algorithmic protocols, adjusting to hosts and external interfaces in their operational format of information exchange.*

The literature review further explores notions of power and control, enacted in networks through their algorithmic aspect. In *When Old Technologies Were New: Thinking about Electric Communication in the Late Nineteenth Century*, Carolyn Marvin explains that "[w]hat was messy was dangerous" (114), elaborating that "[t]o achieve the most widely generalizable abstraction was the goal of scientific knowledge" (112). Admittedly, these two statements are recognized as nineteenth century practices (Marvin 1988); however, they both adhere to the effect that new media has on the general public when addressing networks. Tommaso Venurini, Anders Kristian Munk, and Mathieu Jacomy explain that "the word 'network' . . . can refer to a conceptual topology . . ., to a set of computation techniques (the mathematics of graphs), and to the hypertextual organization of inscriptions (the relational datasets)" (in Vertesi and Ribes 511). Conceptual topology, along with computation techniques, are at the heart of early "communication nets" as defined by Leonard Kleinrock (1964). However, the flexibility of network definitions has arguably allowed for an adaptation of the concept in various respects. That said, networks are unique in their rhythmic nature, along with their functional role in achieving connections between technological devices, people, and institutions across physical spaces. In 1964, Kleinrock defined "Network" as "a finite collection of nodes connected to each other by links" whereby a "Link" is "a one-way communication channel" (12). He also explained that in a network's "Traffic matrix," "the ij entry . . . defines the average number of messages generated per second which have node i as an origin and node j as a destination" (11). In the same context, a "Message" is "specified by its origin, destination, origination time, length, and priority class" (Kleinrock 11). These definitions and explanations highlight the complexities inherent to networks and network design, and showcase that the logic of network operations requires specialization in the field.

Algorithm Studies

The cultural politics of algorithms stem from the fact that "[t]he criteria by which algorithms determine evaluation are obscured while they are enacting political choices about appropriate and legitimate knowledge" (Crawford 86). Algorithmic processes are political by design, which means that the decisions they make and operations they run reflect political biases of their designers and programmers, correspondingly impacting users and the general public. Obscuring criteria from people is not always a decision intended to save audiences from unnecessary information; it is yet another political decision that intentionally masks information likely to contain answers as to how particular actions are calculated, and how goals are achieved. Furthermore, "algorithms can be manipulated in ways that do not disadvantage their users *directly* or *obviously*" (Sandvig et al. 3; my emphasis). This subtle form of manipulation is another political feature that is prominent to the algorithmic culture. Christian Sandvig, Kevin Hamilton, Karrie Karahalios, and Cedric Langbort argue that auditing algorithms should begin by addressing such subtleties, and extend to inspect problematic behaviours that may not arise directly or overtly (Sandvig et al. 3). Kate Crawford reiterates the scholarly argument suggesting that the political nature of technology is "unavoidably linked to institutionalized patterns of power and authority" (78). The connection between the material and institutional politics of networks highlights the important role that people—specifically programmers and engineers— are already playing in this area. This claim does not mean that calling people to participate in questioning cultural politics of networks would be invalid; on the contrary, it highlights their responsibility and emphasizes their role in studying both the technical as well as the social.

In a pragmatic argument, Crawford suggests that scholars need to "go further than simply analyzing the design of the algorithm and pay close attention to shifts in power, from programmers to the algorithms themselves to the wider network of social and material relations" (Crawford 83). Power dynamics within the realm of algorithmic design and operation fluctuate between programmers and the artifact itself—the algorithm—further complicating their effects onto social and material infrastructures. Paying attention to such dynamics directs people to recognize that the cultural politics of algorithms—and by extension networks—are not restricted to their technical structure; rather, they are entwined with the social factors that take part in the formations of such systems. "Advances in the critical self-awareness of technical practice are intellectual contributions in their own right, and they are also necessary conditions for the progress of technical work itself" (Agre 23). Allowing oneself to be more alert to the intricacies of technological complications is in itself an act of participation in algorithmic power dynamics, if not reclaiming one's own power within a dominant system: when people are exploring the cultural politics of networks, with an acknowledgement of the complexities that come into play (social as well as technical), they are starting on the path of a critical and reflective intervention.

While inspecting technical particulars of networks through critically engaging them in everyday life contexts, people are capable of questioning the corresponding cultural politics, comprehending the multiple factors that come into play. Technology today is a domain that allows for user interference or—more realistically—participation, especially since "social actors are the key agents in influencing how [its] features will be shaped over time" (Crawford 78). Sandvig et al. propose an example of action, which would reveal discriminatory behaviour within algorithms, concerning class, race, and gender. The approach they suggest is auditing Internet-based platforms (such as Facebook, YouTube, Netflix, and others), to monitor how algorithms are operating, and to inspect the consequences they may have on social, cultural, and political levels (Sandvig et al. 6). Auditability as an approach requires adding a role that would normally be more or less external to the technical process: researchers, invested users, and people in general are "to hold Internet platforms accountable by routinely auditing them" (Sandvig et al. 18). In that sense, auditing networks is performative: it is critical of the technical behaviour of the system at hand, building off of social and cultural concerns, all the while monitoring it on a regular basis in a way that speaks to the routine element of everyday life, and therefore its rhythms.[9] People who are capable of merging their relationship with their

---

[9] "[T]he routine of everyday life is an emergent phenomenon of moment-to-moment interactions that work out in much the same way from day to day because of the relative stability of our relationship with our environments" (Agre 8).

environments, with the technical and network artifacts available to them, contribute to the larger context within which they exist in ways that can eventually shape "a different narrative and a more diverse cast of political actors" (Crawford 81). That is, when people invest in available technologies to understand their everyday spaces and surroundings, they manage to contribute to the conversation about the various components that politically define their spaces.

In the "Defining Networks" section above, the vocabulary used to define networks and their components invokes notions of time and space—which also appear in the definitions of "Queue", "Queue discipline", and "Message delay" (Kleinrock 11).[10] Henri Lefebvre and Catherine Régulier (in Lefebvre 2013) explain that people "will grasp every being . . ., every entity . . . and every body, both living and non-living, 'symphonically' or 'polyrhythmically'. [They] will grasp it in its space-time, in its place and approximate becoming" (Lefebvre 89). People's relationships with objects/entities around them can be experienced through their rhythms, within the space-time continuum of our interconnections with our surroundings. That said, given the existence of networks within our environments and their operations within our everyday life, we are bound to encounter and even experience them as per their rhythmicality, more so given that "rhythm enters into the lived; though that does not mean it enters into the known" (Lefebvre 86). Consequently, experiencing the rhythms of an infrastructure does not necessarily mean that we directly comprehend—or even apprehend—them. Bringing these rhythms into the known necessitates a representational technique that would translate the non-tangible and inaudible networking processes into a graspable form.

---

[10] Kleinrock (1964) gives the following definitions: a queue is "a waiting line (composed of messages in our case)"; a queue discipline is "a priority rule which determines a message's relative position in the queue"; and a message delay is "the total time that a message spends in the net" (11).

Listening

According to Roland Bathes, "[b]ased on hearing, listening . . . is the very sense of space and of time, by the perception of degrees of remoteness and of regular returns of phonic stimulus" (246). Listening—as well as hearing—allows for people to assess their surroundings, in terms of time and space continuum, through their ability to cognitively process sounds. R. Murray Schafer claims that "[s]ound signals may often be organized into quite elaborate codes permitting messages of considerable complexity to be transmitted to those who can interpret them" (10). That is, sounds can be informational to those who can decipher the signals. Correspondingly, Shannon Mattern explains that "[o]ur tools for urban listening embody particular ways of knowing the city, with implications for how the city is designed, administered, policed, beautified, and maintained" (n.p.). Sound is therefore a material approach to understanding spaces and environments; it permits an engagement with the infrastructures of the everyday life as well as an appreciation of the complexities that networks manifest in their processes.

Pauline Oliveros (2005) proposes that "[t]he depth of listening is related to the expansion of consciousness brought about by inclusive listening. Inclusive listening is impartial, open and receiving and employs global attention. Deep Listening has limitless dimensions" (15). Deep listening is a technique that invites a rather non-discriminatory appreciation or reception of sounds in one's environment. It also brings forth the sounds that one hears, and thus makes them conscious of their sonic surroundings. Oliveros also elaborates on the types of attention that can be involved in listening practices. She explains that "[f]ocal attention, like a lens, produces clear detail limited to the object of attention. Global attention is diffuse and continually expanding to take in the whole of the space/time continuum of sound" (Oliveros 13). Respectively, she claims that "[t]he practice of Deep Listening encourages the balancing of these two forms of attention so that one can flexibly employ both forms and recognize the difference between these two forms of listening" (13). The fact that deep listening engages both focal and global attentions, allows for people to situate the sonic artifact at the centre of their sonic investigation, all the while attending to the general environment within which the listening practice is happening.

Schafer defines "[t]he soundscape [as] any acoustic field of study. We may speak of a musical composition as a soundscape, or a radio program as a soundscape or an acoustic environment as a soundscape" (7). In that sense, sound as an object of study is not necessarily restricted to academic prescriptions. Similarly, Jonathan Sterne explains that "[i]n modern life, sound becomes a problem: an object to be contemplated, reconstructed, and manipulated, something that can be fragmented, industrialized, and bought and sold" (9). Building on this statement, this dissertation resorts to the use of sound as an object for contemplation, fragmentation, and at time manipulation, but more importantly as a medium to communicate a subject of study—networks—otherwise difficult to sense or articulate. Sound Studies addresses listening as a method for engaging with the world, in ways that could make people alert to their surroundings, and actively engaged with their environments. Sterne—in reference to Barry Truax—explains that "as a cultural practice, listening can be an active means of gaining knowledge of a physical environment through the apprehension of variations in sonic character" (96). Listening as a practice—in the context of this dissertation—affords people access to the physical and material nature of various factors in our environment, including technological events and communications.

In *The Audible Past: Cultural Origins of Sound Reproduction* Sterne explains that:

> Listening gets articulated to notions of science, reason, and rationality. Listening
> becomes a technical skill, a skill that can be developed and used toward instrumental
> ends . . . . [A]udition becomes a site through which modern power relations can be
> elaborated, managed, and acted out. Starting in a few select contexts, the very meaning of
> listening drifts toward technical and rational conceptions. Over the long nineteenth
> century, listening becomes a site of skill and potential virtuosity (93).

Listening as an act of participation allows people to access otherwise restricted fields and disciplines, prescribed as technical and inaccessible to the general public; people therefore can redefine notions of power and technical restrictions enclosed within a variety of disciplines. Barthes explains that "[l]istening is henceforth linked (in a thousand varied, indirect forms) to a hermeneutics: to listen is to adopt an attitude of decoding what is obscure, blurred, or mute, to make available to consciousness the 'underside' of meaning (what is experienced, postulated, internationalized as hidden)" (249). Making sense of sounds brings to cognition elements of one's surroundings that otherwise would have remained outside of the recognized constituents of their environment.

In conversation with Miyazaki's work, Mattern points out that "algorithmic sonification [is not] merely a clever means of making computational processes intelligible to non-specialists. Listening has long been an essential skill in computer engineering and programming" (224). While it is true that listening has been used as a diagnostic tool in early computation and engineering practices (Mattern 2018; Miyazaki 2013 and 2015; Sterne 2013;), Miyazaki actually proposes "the use of the term trans-sonic, which denotes the realm of all signals, oscillations, rhythms and flickerings that are inaudible for humans, but can be made – to a certain degree with certain losses – audible again by the use of media technologies such as audio amplification, radio demodulation or software sonification" (n.p.). Sonifying processes that are not originally intended for the human ear, allows for a critical inquiry approach to technologies, in sensible ways that are otherwise not possible. As Miyazaki puts it, "a balanced and appropriate use of all senses might produce the best results for such an epistemic and media archaeological enquiry" (n.p.). As such, listening to networks engages power relations between users and makers of technologies, not only by allowing people an epistemological understanding of a given system, but more importantly by providing them with a technique that makes them cognizant of the system's infrastructure. That is to say, using listening as an active method, people can access aspects of networks that are intentionally kept hidden from the general public.

Listening further allows for "a language of mediation" (Sterne 94), thus opening the experience of talking about a network or even infrastructure to a broader public. Michel Chion (1994) explains that "[w]hen we ask someone to speak about what they have heard, their answers are striking for the heterogeneity of levels of hearing to which they refer. This is because there are at least three modes of listening, each of which addresses different objects" (25). The three modes Chion proposes are "causal listening" (25), "semantic listening" (28), and "reduced listening" (29). These modes of listening engage "conscious and active perception" especially given that "listening, for its part, explores in a field of audition that is given or even imposed on the ear" (Chion 33). Interestingly, listening becomes a technique to access processes otherwise obscured, allowing a form of sensory engagement for which a given system might not have been designed. More so, listening to the sounds of networks also presents as a tactical medium whose "interventions . . . may disturb, [all the while] the outcomes of those disturbances remain uncertain and unpredictable" (Raley 7). Such practices are important in permitting a critical approach to networks, even when they fail as accurate representations of technical intricacies;

listening to networks is in itself informative, if only in making sensible the layers and complexities of network communications streams.

Critical Technical Practice

In *Computation and Human Experience*, Agre urges that "rigorous reflection upon technical ideas and practices becomes an integral part of day-to-day technical work itself" (3). This statement calls upon two important concepts: "reflection" and "technical." Phoebe Sengers, Kirsten Boehner, Shay David, and Joseph 'Jofish' Kaye explain that reflection is the act of "bringing unconscious aspects of experience to conscious awareness, thereby making them available for conscious choice" (Sengers et al. 50). The term technical is the adjective that "relate[s] to any discipline of design that employs mathematical formalization (including any sort of digital design or computer software), whether for engineering design or for scientific modeling" (Agre 17). Given this definition of "technical," it becomes important to recognize the various ways in which networks have become inherent to the contemporary everyday life. The world that we live in and experience today is designed with heavy technical dependencies, be it in transportation sectors, communications, or even entertainment. In that sense, when Agre discusses flaws with Artificial Intelligence (AI) research for instance, he pushes away from the abstract, toward "situated, embodied agents living in the physical world" (4). Correspondingly, he encourages the investment of humans with technology, in a situational way that feeds off of the social and cultural issues of their environment. In doing so, he encourages people to contextually and critically engage with the technical particulars of networks today, to expose their structural politics, and thus contribute to a culturally cognizant advancement of technical practices. Agre states that "[a] primary goal of critical technical work . . . is to cultivate awareness of the assumptions that lie implicit in inherited technical practices" (105). Networks, by nature—because of their functional algorithmic component—embed various social, cultural, and political assumptions, whereby the term "*algorithm* is coming to serve as the name for a particular kind of sociotechnical ensemble, one of a family of systems for knowledge production or decision making" (Gillespie 22). Although some may argue that simply reflecting on a technical matter may reveal its political problem, Agre argues that by practically engaging with technical aspects of the everyday, people are actively participating in the enhancing of such technological artifacts, in ways that are technically and theoretically informed. Instead of merely learning of the issues at stake, a critical technical approach makes them accessible, provides a plan for action, and involves audiences in a collaborative performance of sociotechnical engagement.

In "The Virtues of Critical Technical Practice," Michael Dieter explains that "criticality is an attitude innate to modernity: it is an act of defiance that delimits and transforms existing arrangements of power" (219). Criticality, in this sense, is a state of consistent and recurrent disobedient engagement with issues of power. To critique, however, is to question "the explications through which such distinctions of true and false cohere" (Dieter 219). It is the action that affirms disobedience and approves it through the process of questioning and evaluating. That is, critique is the practice that creates the attitude or state of being critical. Criticality is not an exclusive experience to human beings; it can also define approaches, methodologies, and fields of scholarship—among others.

Critical Making

Critical making is one approach that builds off of criticality to address various issues, with a focus on materiality in relation to the sociotechnical (Buechley and Eisenberg 2008; Dieter 2014; Ratto 2011) and the bodily lived experience (Rajko 2018; Ratto 2011). Matt Ratto explains that the term "critical making" suggests "a desire to theoretically and pragmatically connect two modes of engagement with the world that are often held separate—critical thinking, typically understood as conceptually and linguistically based, and physical 'making,' goal-based material work" (253). Critical making processes bring together scholarly literature (theoretical) and practice (pragmatic), to achieve their analytical ends. In relation to Dieter's definitions, critical making as an approach performs critique of its subject matters—sociotechnical issues and "conceptual space of scholarly knowledge" (Ratto 254). Investing in materiality is one way in which critical making engages the *lived bodily experience* to instigate sociotechnical critique, through conversation and analysis of processes in relevance to corresponding scholarly literature.

Critical making, as Ratto explains, requires three components constituting a project: a literature review; prototype design and making; and iteration, conversation, and reflection (253). These three stages are the process that enables sociotechnical critique—a driving aim for critical making: the first stage highlights the theoretical aspect, the second focuses on the pragmatic, while the third connects the theoretical and the pragmatic. The literature review stage is essential in that it equips the users with the necessary knowledge in order for them to be able to productively engage with the process and perform a critique of the sociotechnical issues at hand. Prototype design and making provides scholars and users with the purely pragmatic that brings forth their lived body experience with the materiality of critical making. It is important to note that though materiality and prototypes are at the core of critical making, they are not merely for the purpose of distribution or consumption, but more to enable "the act of shared construction itself as an activity and a site for enhancing and extending conceptual understandings of critical sociotechnical issues" (Ratto 254). For instance, Leah Buechley and Michael Eisenberg argue that "[i]ntegrating computer science, electrical engineering, textile design, and fashion design, e-textiles cross unusual boundaries, appeal to a broad spectrum of people, and provide novel opportunities for creative experimentation both in engineering and design" (12). The scholars here, for example, suggest a conversation that extends beyond particular users to reach various disciplines and approaches.

The third stage of critical making—iteration, conversation, and reflection—is concerned with steps that allow users to engage with both the technology and the theory, "wrestling with the technical prototypes, exploring the various configurations and alternative possibilities, and using them to express, critique, and extend relevant concepts, theories, and models" (Ratto 253). Dieter explains that "by decoupling sociotechnical ensembles from seemingly stable configurations, [the physical diagram] prompts and propels a consideration of how things might be otherwise" (226). Working with material products—whether prototypes or finished products—to explore possibilities, exposes the users to the targeted technical and social issues. Dieter suggests that "complex questions of material participation [can] transform frames of technology, while enacting topological renditions of sociotechnical problematics" (224). When users interact with material equipment, for example, and reflect on their experience in light of the sociotechnical environment of the work, they are drawing the connection between "the lived space of the body and the conceptual space of scholarly knowledge" (254). Furthermore, participating in complex conversations about the materiality of prototypes confirms "the constructive process as the site for analysis" (Ratto 253). Designing and making, for example, are key processes during which people can engage in critical thought, especially when they are interacting with physical artifacts deeply situated in the sociotechnical everyday.

Buechley and Eisenberg conduct what can be an example of a critical making project, from the perspective of the organizers (though not necessarily users or learners). The scholars create the LilyPad Arduino kit that allows users—of various levels of expertise—"to build soft, wearable computers" (12). The aim of this kit is to allow users to experiment with electronic textiles (design, engineer, and build).[11] The experience that Buechley and Eisenberg create, highlights the second stage of a critical making project that Ratto proposes—prototype design and making—in that the users design and build their own prototypes. To "evaluate [their] kit, [they] conducted user studies that have interesting implications for pervasive computing, engineering, education, and the technology world more broadly" (Buechley and Eisenberg 12-3). Though the authors do not explicitly discuss the literature review stage that has been conducted as a first stage, they do address it in the design of their study. Based on the workshops that they offered and led, they made iterations, and engaged in an exploration of complex issues, including "motivational and affective issues" (Buechley and Eisenberg 14). The published work resulting from the studies is a contribution to critical making "through the sharing of results [for] an ongoing critical analysis of materials, designs, constraints, and outcomes" (Ratto 253). Despite the intention of making such knowledge public, publishing results does not guarantee that they be shared with the general public, especially because of the language and vocabulary that targets scholarly and academic circles.

Ratto argues that there is a "disconnect between conceptual understandings of technological objects and our material experiences with them" (Ratto 253), which in itself may result in social, political, and technical issues. While people may have a conceptual understanding of the risks and problems with networked technological devices, they do not necessarily engage with them in a critical way that actually addresses their concerns. Critical making aims at exploring the everyday lived body (embodied) experience to unpack the

---

[11] E-textiles are "soft, fabric-based computers" (Buechley and Eisenberg 12).

"everyday sociotechnical arrangements [that] are presented as a problematic object to thought"
(Dieter 224). Although one may think of embodiment as critique, it is in fact critical, because as
Jessica Rajko explain, "embodiment is not something one achieves merely through observation
and design. Embodiment is a *way of being*" (198). Additionally, Ratto claims that his "goal is to
make concepts more apprehendable, to bring them in ways to the body, not only the brain" (254).
Encouraging a physical and bodily experience with objects of study builds on the everyday life
experience, because it allows one to critique the subject matter in relevance to the common and
well-known aspects of human agency. The embodied experience of materiality allows access to
sociotechnical phenomena, not only by drawing on new technologies, but mostly through "re-
experiencing familiar spaces, actions, and interactions over and over again as if they were new"
(Rajko 199).[12]

Engaging with technologies in a known space-time continuum, with attention to the
embodied aspects of the experience allows people a fresh perspective of their environment, and
more importantly of the relationship between themselves, the artifact they are experiencing, and
the environment within which they are existing during that moment. This is not to say that new
technologies cannot enable embodied experiences that achieve critical objectives. Critical
making invests in producing such technologies and projects, drawing on Seymour Papert's
"transitional objects," in that they are "a way of connecting the sensorimotor 'body knowledge'
of a learner to more abstract understandings" and "act as means for projecting oneself into an
abstraction" (Ratto 254). That is, experimenting with computers and other technologies through
the lived body experience pragmatically relates the user to the theoretical aims of critical

---

[12] Consider for example re-experiencing spaces with more attention to their soundscape, each time with a
focus on a certain element over the other—Chapter 4 elaborates on such experiences.

making, and renders more concrete the issues that otherwise would not be physically and materially accessible. In other words, critical making aims at rendering tangible the theoretical realm of its literature review stage, so that discussions and conversations become more accessible. Dieter reiterates this form of experience, highlighting the importance of "experiencing the abstraction of digital systems through encounters that disturb the formal universalities of communication and exchange" (225).[13] That is, he suggests that engaging with networks outside of the normalized experiences would allow people to challenge the way these systems act within their everyday life; such encounters align with Marshall McLuhan's concept of anti-environments, as disruptions of the afore mentioned "formal universalities."

In an example and based on a theoretical overview that he provided, Ratto explains that in the experiment Flwr Pwr, participants were allowed to interfere with, and change the code of their prototypes, to make them communicate in different ways (257).[14] In the discussions that followed the practical stage of the workshop, participants recognized social and behavioural consequences of the various decisions involved in the tinkering process (Ratto 257). Experimenting with the digital by physically interacting with the prototype—tinkering, observing, listening, and subsequently conversing—allows users to be immersed in the experience in a way that connects their everyday social experiences to the theoretical and pragmatic components of their work. Through its focus on conversation, critical making reveals complex structures that would otherwise remain abstract or supressed, thus opening the "conceptual space of scholarly knowledge to a wider audience" (Ratto 254).

---

[13] In that sense, listening to abstract and otherwise inaudible events, such as networks, is a contribution to the disturbance of universalised "black-boxed" networks, as will be reviewed below.

[14] Flwr Pwr is a workshop that Ratto organized "to create a shared construction exercise that could facilitate and inform discussions around the rise of proprietary and closed 'walled gardens' on the Internet and provide some common ground for thinking through the social issues involved" (255).

Although critical making has potential in that it brings together material and theoretical facets of technological and social issues, it still has some limitations that one needs to consider when planning a project. Ratto explains that critical making risks a lack of connection between object and topic in question, which results in a disconnect between the material and conceptual, and interferes with the potential productive analytical discussion. Dieter also explains that "poorly understood technical matters can easily lead to unintended or unanticipated outcomes" (223). Assuming knowledge with regards to a given technological issue is likely to complicate the discussion, interrupting the flow of critical engagement. For instance, in one of Ratto's projects, one complication was that "the majority of [the participants] had trouble mapping what they had just done to the critical issues involved" (Ratto 255).[15] When participants cannot relate the pragmatic to the theoretical, the critical conversation cannot take place, and if it does, it will not be quite productive.[16] One way to address the concern of poor understanding is to have a balance "between technical and social scholarly expertise" (Ratto 255); it is essential to understand the social or cultural issues that people are trying to question, before engaging in a material and technical experiment as a critical performance. Additionally, participants may not be interested or invested in the subject of study, which can in turn disrupt the process and create difficulty in connecting materiality, lived bodily experience, and the theoretical context (Ratto 255). Dieter argues that "operationalized modeling and logical processing of computation generate ambiguities, paradoxes, and other expressions also handled as *exceptions*" (225). Such a statement applies to the lack of participant investment (resulting in ambiguities and disruptions),

---

[15] RCA/Imperial Event: Systems of Learning (Ratto 254-5).
[16] Some may argue that pointing out the disruption discussed above can be in itself a critical intervention to the progress of a given project. Consider Dieter's claim: "orchestration of disruption itself delivers significant redistributions of power, wealth, and influence" (223).

and extends to external audiences. Ratto remarks that "it remained (and remains) quite easy for external viewers to misunderstand either the intention of the exercise or the relevance of its results without extended commentary such as this article" (255). This limitation suggests that a critical making work is not necessarily self-sufficient, but requires an extensive scholarly support that bridges between various scholarly communities.

Critical Design

Critical making calls for projects that combine theory, design, and making with human agency and embodied (lived body) experience. The result of this combination is a form of critical conversation that aims at joining various groups and communities, addressing sociotechnical issues and rendering their critique more accessible. It engages scholars and users in a conceptual space that addresses Bruno Latour's notion of "matter of fact" versus "matter of concern'" (2004), which Ratto, in turn, takes on to be at the heart of critical making: it "is about turning the relationship between technology and society from a 'matter of fact' into a 'matter of concern'" (259). In other of words, instead of accepting that technologies work the way they do and there is not much room for interventions, people are invited to question the designs of technological operations and how their network functionalities manifest in our everyday life. One can thus argue that critical design and making, through material experimentation and analytical conversation, promotes a shift from the instrumentalist and deterministic understanding of technology to one that is critically experienced and lived. Simon Penny argues that designing experiences benefits from artistic methodology. To design systems that allow for new experiences, Sengers explains that "one must consider the technical challenges to be overcome in the context of the kinds of *cultural* and *social* meaning that the system may take on and the ways in which *users* may *choose to interact* with it" (20; my emphasis). Systems have two facets when it comes to social and cultural implications: one is inherent to the system design itself, while the other corresponds to how people interact with the system in a way that is influenced by their own social and cultural backgrounds. When designing experience, the focus begins with the user, since they have the agency to choose how to interact with the system.

The cultural and social environments of the user condition the outcome of the system's design, because "[m]eaning is construed entirely as a result of the observers' cultural training" (Penny n.p). When interacting with or experiencing a system, "[w]e extrapolate from our previous experience to explain our new perceptions" (Penny n.p.). For example, people's experience of a given environment changes after they hear the sounds of networks, though these experiences would still be deeply associated with the cultural and social meanings that the new sounds might carry. Designing with a cultural inclination and social acknowledgment highlights the experience of people and enhances their interaction with the designed objects in question, allowing for "thoughtful artefacts that create opportunities for thinking about and engaging in new kinds of experiences" (Sengers 21). The cultural and social environments of the users naturally dictate their behaviour, which could subsequently be interpreted by a machine system—networked and algorithmic as it may be—thus dictating its behaviour. The importance of the user experience here highlights the cultural relevance in designing experience, especially when/if "[t]he goal is to establish a metaphorical interactive order where the user's movement 'corresponds' to some permutation of the output" (Penny n.p.). When the relationship between the artifact and the user is interactive and responsive, the cultural and social dynamics between the two is exacerbated. Accordingly, the system must appeal to the user and engage them continually, for a more productive system (Penny n.p.). Designing experience *assumes* a level of engagement on the part of the user that dictates the functionality of the system; the design process and the user-interactive process thus require delicate and calculated attention to maintain a successfully engaging system.

The design processes must acknowledge questions of cultural and social significance of particular technologies and interfaces. Whether design is concerned with objects or experiences, it still considers significant social aspects, and addresses values that range between the cultural and the political. The political, as Chantal Mouffe explains, is "the dimension of antagonism that is inherent in human relations, antagonism that can take many forms and emerge in different types of social relations" (101). Carl DiSalvo elaborates on this definition, explaining that "the political is a *condition* of life—a condition of ongoing contest between forces or ideals" that "can also be expressed and experienced through design" (8). That is to say, the political requires people to interact with external artifacts, which is where design can contribute to enhancing and directing their experience. Daniela K. Rosner and Morgan G. Ames speak of *"negotiated endurance,"* which is "the process by which different actors – including consumers, community organizers, and others – drive their ongoing use, maintenance, and repair of a given technology through the sociocultural and socioeconomic infrastructures they inhabit and produce" (319). Negotiated endurance highlights the importance of the political—as a condition of life immersed within the sociocultural and socioeconomic—giving people the power to interact with design projects, with an awareness not limited to the mere function of objects at hand. On the other hand, Phoebe Sengers calls for "designing [experiences], using principles that draw on both technology design methods and social and cultural analysis" (21). That is, she is calling for an integration of cultural issues in the designing process, to create (user) experiences that will allow people to culturally and socially engage and interact with the designed element. Although designing objects assumes political—and socioeconomic—values, while designing experiences builds on cultural values, both approaches rely on the user's interaction with the product to achieve or reveal the values it assumes.[17]

A Brief Response

This dissertation mainly focuses on the mechanisms of networks as established through algorithms and EM waves mechanisms, particularly when translated into communication events. Further, it acknowledges that, as they stand to the average user, networks seem to be chaotic, despite the mathematical and logical characteristics attributed to them in their protocols and behaviours. As this project explores, network stories most often associate subject networks with a sense of fear, mystery, and pathology. The chaotic and multidimensional aspect of networks is often responsible for inducing fear in audiences, and thus reinforcing the control that networks are capable of precipitating. Additionally, the abstraction of networks into mysterious and undefined systems broadens the gap between the general public and computer scientists and engineers, thus unintentionally—and perhaps unknowingly—exacerbating the power dynamic prominent in the sociotechnical realm of everyday life. That is, the gap between technology designers and makers and technology users is aggravated by the fact that users do not necessarily—want to—know how networks operate. In fact, the mystery factor is part of the design, especially in that it contributes to the convenience of use: people do not need to trouble themselves with the technicalities of technologies in order to use them; the less they know about technical complexities, the easier and more pleasant their experiences can be.

Network communications in this day and age are at the heart of people's everyday life, even if we still talk about networks in abstract terms. Networks, being the infrastructure that

---

[17] This is not to say that designing objects cannot be cultural, nor does it keep the political away from designing experience.

makes our digital dependencies possible, remain ambiguous particularly because the term "network" has been conflated to refer to various concepts including but not limited to: visualizations, social connections, and digital social frameworks. In that sense, approaching networks requires a technique designed to allow people to appreciate the intrinsic complexities of such systems, acknowledging their messy layers and admitting their inaccessibility. While STS scholarship focuses on investigating technologies mostly from a personal computing perspective (Buechley and Eisenberg 2008; Dieter 2014; Ratto 2011; Rosner and Ames 2014), few scholars specifically address networks as a subject of study (Miyazaki 2013 and 2016). Although it might be difficult to address this topic with clear and direct linguistic style, it remains necessary to find ways to present it to the general audience using an approach that is not too alienating. The political and sociotechnical placement of networks requires a level of awareness on behalf of people who use and participate in them, which cannot be achieved when network stories are as abstract and ambiguous as the networks themselves. In other words, people should have access to network stories that situate the subject network within a corresponding familiar environment. These stories would not only allow people to realize the interconnection between networks and everyday spaces, but more importantly echo the structures of politics and control inextricable from technological invasive systems.

In an attempt to mitigate these issues, this dissertation resorts to the use of sound as an object for contemplation, fragmentation, and at time manipulation, but more importantly as a medium to communicate a subject of study—networks—otherwise difficult to sense or articulate. As such, designing a technique that exacerbates the power that listening grants the general public, is a contribution to the already existing critical practices challenging the power and control features of infrastructures, as demonstrated in Jacqueline Wernimont's *Living Net* (2018).

While listening allows for sensory and thus embodied experiences, such a technique might be subject to Ratto's concerns with requiring further conversation for more robust understanding of the critical process. However, in engaging listening to networks as a tactical medium, one allows for a flexibility that escapes "fixed or exclusive" critiques (Raley 6).

Conclusion

Science and Technology Studies provide an array of scholarly work that engages methods for intervening in the techno-culture of everyday life. This dissertation builds on existing scholarship, to evaluate how critical and material encounters with technologies can be used in telling network stories, making them accessible to people outside the giant-tech sphere. Networks are complicated systems with multiple possible definitions. Aside from their metaphorical and visual dimensions, networks are commonly known for their infrastructural nature and connectivity functions. At the core of these systems, algorithms manifest as regulatory and controlling components of network infrastructure, seeping into the social through the mere use of network technologies. For example, people seek or occasionally actively avoid technological connectivity—through Wi-Fi or GSM—and at times travel with these concerns at the forefront of their itinerary. Algorithms, rhythmic in nature, are designed to silently and calmly achieve their functions. However, given their materially established through EM waves, people may attempt to listen to their operations and rhythms. Listening is an approach that connects people to their surroundings, and attunes them to the (sonic) nature of their environments. Sonification also function as a diagnostic technique that could alert the listener to the status of the subject of study. Similarly, critical making and design acknowledge the complexities of technologies and invite processes of reflection and theoretical grounding, especially through embodied experience. Tactical media promises flexible and engaging outcomes that place audiences at the centre of the projected outcome. It does not determine a specific or restricted product, but aims at disturbances that provoke thought and questioning. Similar to critical making and critical design (mostly through design for experiences), tactical media engages materiality to appeal to embodied and lived experiences.

Networks are subject to critical inquiry given their complex and highly technical nature. However, networks are sheltered from our sensory motors through their wireless and EM connections. Addressing networks, their infrastructures, and their processes relies heavily on the limited tangible components that constitute them, most apparent of which are algorithms. Given that algorithms are mathematical and computational commands that people author, the role of humans in network processes remains central, yet complicates how we address them. In telling stories about networks, people set predetermined goals in anticipation of audience reception. The complexity of networks enforces a storytelling approach that either glorifies the network sphere, or renders it magical. On occasions, addressing network failures pathologizes the infrastructure; one could argue that such a language relates to the "medical epistemology of pathological anatomy" (Sterne 103). Similarly, the fact that networks and algorithms are inherently rhythmic—through their time-space features, traffic matrices, spatio-temporal delays (Kleinrock 1964)—pathological rhythms can also manifest as means to evaluate technological functionalities. In that sense, approaching networks would benefit from bringing forth their rhythmic processes, without necessarily pathologizing them. Notably, a network story that is designed with predetermined audience interpretation risks misinformation and disinformation, if only in presenting it as a collection of magical processes, even though they are actually materially established. With that in mind, this dissertation brings together methods from various branches of Science and Technology Studies; it investigates listening techniques, applies critical design and making, and bridges audience-oriented designing for experience methodologies with the provocative and disruptive tactical media approach.

Chapter 2: Rhythms and Network Stories

Various types of media define, address, and investigate network stories, including

technical (and not so technical) definitions (Advanced Research Project Agency Network

[ARPANet]), news stories and snippets (American Telephone & Telegraph Corporation [AT&T]

1990 crash, and Toronto Smart City), as well as TV series (*Halt and Catch Fire*). As defined in

the introduction, network stories are *stories that people tell about networks, network*

communications, and technologies that rely on networks to function. Using rhetorical analysis

and close reading, relying on the science and technology studies (STS) theoretical framework,

people can better understand the meaning and effects of the stories in question, thus addressing

the tangible effects of networks, as hypothesised or reported by the media. Henri Lefebvre

(2013) argues that "[a] skillfully utilised and technicised form of mythification (simplification),

[imagery] resembles the real and presence as a photo of photographed people . . . . Yet the

image, as the present, takes care of ideology: it contains it and masks it" (56). Ambiguities

around networks are often rooted in a calculated attempt for maintaining the magical aspects

(and mystification) of network communications. In such cases, the media portrays and distributes

information about technology in general, and networks in particular, in a way that asserts power

and control: "the output (rhythm) changes according to intention and the hour" (Lefebvre 57).

However, it is not always the case that media intentionally uses ambiguation to

participate in a power dynamic, characteristic of techno-culture; ambiguation is at times

necessary, to avoid distracting audiences with complex information that require certain expertise

to comprehend. While media engages with technical details at varied levels, extreme

representations (either utopian or disastrous) seem to be a common characteristic. Stories around

networks often serve a biased role that under-informs audiences, imposing an acceptance of technologies as magic, where the unknown is to be either feared and boycotted, or praised and invested in. Either way, audiences are frequently expected to blindly and faithfully accept what is being advertised, via "uncritical absorptions of the medium's self—or seemingly self—evident representations" (Kirschenbaum 36). Matthew Kirschenbaum explains that medial ideology "substitutes popular representations of a medium, socially constructed and culturally activated to perform specific kinds of works, for a more comprehensive treatment of the material particulars of a given technology" (36). The medium that communicates network stories thus matters in that it determines how the message ought to be perceived, and how the story should be told, particularly because "the media day unfolds, polyrhythmically" (Lefebvre 57). The social and cultural implications of digital journalism are different than those of TV and film, because different media serve different purposes, and thus adopt different rhythms. Nonetheless, different media remain equally noteworthy in discussions around network storytelling, given the unique particulars of their rhythms and mechanisms, how each approaches the event in question, and the technicalities involved in the storytelling processes. In this chapter, I argue that network stories have rhythms that represent those informed by technologically communicated human actions, consequently reflecting the abstract spaces of networks.

One of the earliest networking projects, the (publicly known) first of them, is the ARPANet. The project arguably started in 1966 (or "began development" as per Computer Hope) or 1969 ("conceptualized," according to TechTarget - SearchNetworking); however, Live Science claims that its conceptualization dates back to 1961 in a paper by Leonard Kleinrock. Computerhope.com defines ARPANet as "a Wide Area Network linking many universities and research centres, [which] was the first to use packet switching, and was the beginning of what we consider the Internet today" (n.p.). This definition highlights the importance of ARPANet and its role as the precursor for the "Internet," suggesting that it is the foremost reason behind the connection of the world's computers. On the other hand, techopedia.com explains that ARPANet "was funded by the [United States] government during the cold war to build a robust and reliable communications network" (n.p.). As per the definition, the network would connect "various computers that could simultaneously communicate in a network that would not go down and continue running when a single node was taken out" (techopedia, n.p.). The connection to the war is important for two reasons: first, it associates ARPANet with a kind of violence, suggesting that telecommunications for that matter started in service of warfare; second, it emphasizes that the reason why this network was highly important is its stability and resilience to particular failures and disconnections in critical moments (on a national level). In turn, this connection participates in the assumption that telecommunications today must be "magically" resistant to all kinds of interferences and disruptions.

According to TechTarget - SearchNetworking:

[ARPANet's] initial purpose was to communicate with and share computer resources among mainly scientific users at the connected institutions. ARPANet took advantage of the new idea of sending information in small units called packets that could be routed on different paths and reconstructed at their destination. The development of [Transmission Control Protocol/Internet Protocol] TCP/IP protocols in the 1970s made it possible to expand the size of the network, which now had become a network of networks, in an orderly way (n.p.).

Highlighting the importance of scientific users and their centrality to ARPANet's purpose supports the idea that technology is (mainly) designed in support of scientific work, to advance such efforts and enhance connectivity amongst researchers in particular institutions. This notion in turn suggests a certain hierarchy and privilege that is still often associated with hi-tech. On the other hand, bringing to attention that TCP/IP protocols allowed for an increase in the size of ARPANet in an "orderly way" is a factor that mostly stresses how protocols allow for, if not ensure, order and control in large and complicated systems. More so, this notion coincides with Alexander Galloway's claim that "protocol is a distributed management system that allows control to exist within a heterogeneous material milieu" (8). That is to say, protocols allow control in environments that would otherwise not be subject to the same laws, given their inherent differences and material independences.

Further, LiveScience connects ARPANet's history to the World Wide Web (WWW), crediting its concept to Kleinrock, explaining that "he wrote about ARPANet, the predecessor of the Internet, in a paper entitled 'Information Flow in Large Communication Nets'" (Zimmerman and Emspak n.p.). Relating ARPANet to the WWW alludes to a sense of specificity, in that the WWW may be perceived as less mysterious than the "Internet." Furthermore, using Kleinrock's name and the title of his paper adds credibility to the source, providing the users with a reference that they can consult if curious. The website also draws connections between ARPANet and ubiquitous telecommunications including e-mail, Facebook, and Twitter, claiming that it is the "backbone" of such current platforms; it gives readers a relevant and contemporary reference that can assist them in understanding the general concept. Nonetheless, this move reiterates the magical hints often associated with networks, given the lack of detail around the connection suggested or the developmental process in question. The rhythmic flows of the different stories around ARPANet vary in their time-space flows, where each story focuses on a certain time or space: ComputerHope situates ARPANet in the past, with a connection to the present (though static) state of the Internet; techopedia associates the network with the cold war, and by extension links to the rhythms of war and violence; TechTarget – SearchNetworking emphasizes the importance of the protocological component of the network, thus reinforcing a rhythm of control; and LiveScience combines past and present—similarly to ComputerHope—in connecting ARPANet to the WWW, though it does so in a way that brings forth the contemporary rhythms of social media that have different flows than those of the more static WWW (at least in the context of its creation as referenced in the ComputerHope story). Such distinctions correspond to Dawn Lyon's (2019) "nuanced understanding of the articulations of tempo, movement, flow, stasis and repetition" (4). While these stories are contextualized as definitions of a network, each carries its own rhythms and associations to temporality and sociocultural spaces.

Technology also manifests in the media via news snippets and breaking stories. Network failures most commonly attract media attention, allowing for stories to be told in multiple ways, some of which supporting technological advances, while others intentionally (and perhaps dramatically) calling for fear and opposition. On January 15th, 1990, the AT&T company witnessed what is known as a bug—a problem in the coding/programming of a software—that crashed its long-distance network for nine hours (Miyazaki 133). Upon the implementation of a software update in centres across the United States, a shutdown of one of the networked machines led neighbouring machines to initiate self-shutdowns as they rerouted their network maps: a problematic and disruptive feedback loop. According to Shintaro Miyazaki, the incident was "popularized by Bruce Sterling, a science fiction author and pioneer of digital cultures, in *The Hacker Crackdown*, his first nonfiction book published in 1992" (133). Journal articles also covered the event, providing audiences with their first access to the details of the story. It is not surprising that the anger and frustration of people affected by this disruption, translates to a more or less radical dissemination of fear by questioning the stability of networks in general and AT&T in particular. The event, after all, was a "polyrhythmic refrain" (Miyazaki 133), which coincides with media operations, but more importantly imposes a sense of arrhythmia when stories are not fully coherent with the event itself.

While the AT&T crash stories report a technical failure that already manifested in financial and business damages, contemporary tech stories tend to anticipate disruption and speculate complications. Toronto Smart City is a contemporary network project that was very present in the digital news media realm, until it was shut down in February 2020. In collaboration with Waterfront Toronto, Google sister company Sidewalk Labs planned a smart city that extends on twelve acres of the Toronto waterfront. As a precursor, and a test project to Toronto's Smart City, Sidewalk built a prototype in Quayside, a "planned pilot location for testing what could be implemented on a larger scale" (Kotecki n.p.). The project promotes many urban advances, including but not limited to: modular housing, self-driving shuttles, and renewable energy systems (Kotecki 2018: n.p.). The idea, as advertised by the site's authorities, suggests a use of technologies to serve the envisioned community, which explains why Waterfront Toronto has received support from local and national governments, estimated to amount to $996 million (Kotecki n.p.). In parallel, Waterfront Toronto's CEO promoted the use of "smart" technology (also known as Artificial Intelligence) in a way that aligns with Mark Weiser and John Seely Brown's claim that calm technology would enrich "our opportunities for being with other people" (85). The dissemination of information and data reports around this project was distributed across various fronts, including the Sidewalk Toronto website, and the various news websites—such as British Broadcasting Corporation (BBC) or Canadian Broadcasting Corporation (CBC), and The Guardian (to name a few). As opposed to promises made around investing in smart technology in service of the community, the privacy expert of the project, Dr. Ann Cavoukian, resigned in October 2018, stating that she "imagined [the project leaders] creating a Smart City of Privacy, as opposed to a Smart City of Surveillance" (Canon n.p.).

Furthermore, Saadia Muzaffar, a member of the digital strategy advisory panel also stepped down, as "the company was not adequately addressing privacy issues she and others had raised" (Canon n.p.). According to The Guardian, Alyssa Harvey Dawson, head of data governance, explained that Sidewalk projects "would 'set a new model for responsible data used in cities', but that the company was still working to settle a plan" (Canon n.p.). By introducing the term "responsible data," Dawson delivers positive connotations and promises, even when the terms are not completely defined or explained. In circulating such a technical term in the media, she not only raises legitimate and credible concerns among audiences, but also gives them the language to investigate, question, and discuss these fears. Further, she suggests that "entities proposing to collect or use urban data (including Sidewalk Labs) will have to file a Responsible Data Impact Assessment with the Data Trust that is publicly available and reviewable" (Canon n.p.). This statement, however, is problematic in various ways. First, the article does not clearly explain what the Data Trust is, who would be responsible for it, and what data it would hold. Second, the requirements that need to be included in a Responsible Data Impact Assessment are not mentioned nor explained. Third, this statement claims that the Data Trust is publicly available and reviewable, which is a vague notion that can suggest a variety of complications, given that the public (i.e. anyone) will have the ability to view and review the data.

One may argue that the use of data itself is problematic, but in terms of privacy, any access to people's information should be conditional and restricted. It is important to note that

Sidewalk is arguing that the company's role in implementing privacy by design principles will be limited, as they will not be able to control how other companies involved in the project will handle data (Canon n.p.). Such a disclaimer gives them a window to escape potential complicated scenarios when, later in the life of the project, data gets misused or abused.

Sidewalk also uses this statement to explain the reasoning behind Cavoukian's resignation. Sidewalk gave up on the project in May 2020, when CEO Dan Doctoroff stated that

as unprecedented economic uncertainty has set in around the world and in the Toronto real estate market, it has become too difficult to make the 12-acre project financially viable without sacrificing core parts of the plan we had developed together with Waterfront Toronto to build a truly inclusive, sustainable community (n.p.).

The polyrhythms of the Quayside story manifest in the various tones of the various parties involved—ranging from positive and promising announcements to sceptical descriptions and resignation statements. These polyrhythms raise multiple flags, particularly because of the different messages between the CEO's stories and those reflecting issues of privacy and surveillance. These multiple—and for the most part conflicting—rhythms oppose Lefebvre's notion of **"mediatised everyday"**: "The repetitive monotony of the everyday, rhythmed by the (mediatised) media need not bring about the forgetting of the exceptional" (59). The rhythms of Toronto Smart City network story conflate and exclude the dialogue replacing it with the mundane and passive subject (Lefebvre 2013): the supposedly exceptional stances that Cavoukian and Muzaffar take are reduced to a "banal … publicity label: 'Here is the exceptional'" (Lefebvre 59), without engaging the events in rhythmic dialogues. While the acts themselves employ strong rhythms, the stories merge them into the polyrhythmicality of media, discrediting the tempo of exceptional acts, for that of the controlled narrative of Sidewalk Labs.

In an extreme contradiction to the ignored rhythms of exceptional "mediatised everyday" (Lefebvre 59), visual shows and films present themselves as another form of media that discusses networks, with an intentional focus on—if not inflation of—the exceptional in network stories. *Halt and Catch Fire* is an acclaimed TV series that explores the lives of the people behind technological innovations, through a fictional and romanticised lens. According to The Guardian, this show is a "tech drama that takes place entirely between the first iteration of Microsoft Word in 1983 and Windows 95 . . ., [focusing] squarely on the haze of an emerging field" (Thurm n.p.). This TV show embraces the life of characters who namely pushed the technological industry forward, by attempting innovative moves, yet failing as they compete with bigger and more dominant companies. In its last season, *Halt and Catch Fire* focuses on the process behind an expansion and commercialization of a network based on the National Science Foundation Network (NSFNet), via access to ARPANet credentials. Although many of the companies, software, and ideas presented in the show are fictional (Thurm n.p.), the allusions and relationships drawn are explicit, if not intentionally emphasized, suggesting a different medium for, and approach to, network stories. The rhythmicality of the show is remarkable in that it spans over an extended period of time, with delicate attention to the flow and tempo of the character's lives, in accordance to those of the fast-paced technological advances. The polyrhythmicality of the show is arguably an eurhythmia, in that the rhythms of characters' growth and development match up to those of technological advancements, all the while acknowledging the clear break between what is feasible in terms of personal growth as opposed to what is achievable in the realm of technological advancements. The next section further explores rhythms of the network stories introduced above.

Rhythms of Network Stories

Lefebvre emphasizes the importance of the study "of the modification of [the natural] rhythms and their inscription in space by means of human actions, especially work-related actions" (117). In the age of the Internet of Things (IoT), many—if not most—of the human work-related actions are performed, documented, or achieved within a network infrastructure. Lefebvre continues to explain that "[t]raversed now by pathways and patterned by networks, natural space changes: one might say that practical activity writes upon nature albeit in a scrawling hand, and that this writing implies a particular representation of space" (117). Technological networks constitute an abstract space, written upon by algorithms and human commands, to be then represented in stories through various media. Stories about these systems thus embed multiple layers of rhythms: they embody the rhythms of the event they address and reflect contextual rhythms—political and social. According to Lefebvre, "[rhythm] is not a thing, nor an aggregation of things, nor yet a simple flow. It embodies its own law, its own regularity, which it derives from space – from its own space – and from a relationship between space and time" (206). Within the context of network stories, it is important to note that the space varies. If the space/space-time continuum in question is that of the event, then the rhythm in question would be that embodied by the story. However, if the space/space-time continuum is that of audiences or publication, then the rhythms in question are those reflecting the contextual rhythms. Nonetheless, "[r]hythms in all their multiplicity interpenetrate one another" (Lefebvre 205), which suggests the possibility—if not need—to not only approach these rhythms separately, but also as the rhythms of one body, or infrastructure for that matter. Although network stories have different rhythms, these are interconnected and can represent a single body of rhythms for the event as a system. Studying network stories allows people to inspect the representation of abstract network spaces, and how they reshape natural space through the multiple rhythms of technologically communicated human actions.

The ARPANet stories present the network to the public as one mostly rooted in the military context that greatly contributed to the growth of the network. As per Techopedia's definition, the government funded ARPANet for the purpose of creating a network that does not crash if one of its nodes was down (techopedia, n.d). Notably, the government supported the network so as to maintain a eurhythmia, even when a discordance takes place.[18] Further, ARPANet definitions also situate the network within a military context, explaining that "[i]n the 1980's, ARPANet was handed over to a separate new military network, the Defence Data Network, and NSFNet, a network of scientific and academic computers funded by the National Security Foundation" (TechTarget – SearchNetworking, n.d.). This aspect of the ARPANet particularly coincides with Lefebvre's claim that

> Societies marked by the military model preserve and extend this rhythm through all phases of our temporality: repetition pushed to the point of automatism and the memorisation of gestures—differences, some foreseen and expected, other unexpected— the element of the unforeseen! Wouldn't this be the secret of the magic of the periodisations at the heart of the everyday? (48).

Societies that have infrastructures rooted in military funding and modeling acquire rhythms that redefine social temporalities; these rhythms replay automated sequences of "magical" operations, reinforcing themselves as integral factors in the everyday. Exploring ARPANet as a social element (one that has had and continues to have rippling effects on the everyday), expands our

---

[18] Lefebvre (2013) defines eurhythmia as "Rhythms unite with one another in the state of health, in normal . . . everydayness" (25).

understanding of it from a rhythmanalysis perspective. The rhythmic setting of this network is not only within its composition, but also an important part of its story. In other words, the growth of the network and its expansion have a repetitive pattern that nonetheless embodies differences (such as the switch into the Defense Network and the NSFNet). Further, the move to NSFNet (in 1990) happened after a complete technical transition to TCP/IP (in 1983) (ComputerHope n.d.). The technological advancement that allowed for the growth of the network is the "*secret* of the magic" (Lefebvre 48; my emphasis) element (the switch to TCP/IP), that marked the beginning of the new period for networks. The rhythms of this network resulted in remarkable imprints of human actions onto the natural space, beginning to redefine the natural space so as to encompass people's technological prints.

In *Halt and Catch Fire*, Joe MacMillan applies the "magic" in his visions of "the next thing" when brainstorming with Ryan Ray. He recognizes the power of his time period: openness (S3:E4). In their discussions, MacMillan and Ray do not necessarily speak of the great technologies that are available to them (outside of a quick listing on the fourth episode of third season). The conversation between the two characters is an example of the magic in technology, as it merely speaks of the specifics, yet glorifies openness—a notion that is known to the general public, though its components and how it came to be, remain mysterious. MacMillan's talk of openness ("Glasnost, it means openness. And it's always a bad idea to bet against openness") is also a utilization of two specific types of rhythms, as defined by Lefebvre: "*Fictional rhythms*: Eloquence and verbal rhythms, but also elegance, gestures and learning processes . . . . The imaginary!" and "*Dominating-dominated rhythms*: Completely made up: everyday or long-lasting, in music or in speech, aiming for an effect that is beyond themselves" (27). In the context of this scene, the dominating-dominated rhythm is within the more encompassing fictional rhythm, which is that of the show as a whole. However, the design of this particular speech aims to leave an impact on audiences, pushing them to support and encourage MacMillan's newest endeavour, despite the other failures and troubles he had caused upon other characters. These rhythms thus work together in order to create a story that is accessible to audiences; they leave them with enough yet minimal technical substance, but establish ideas around the technological progress that shapes the everyday life of the characters, and more importantly, is assumed to have shaped the everyday of audiences as well.

Contrastingly, pathologies exist within rhythms as they do within other aspects of the everyday life. Lefebvre explains that "when [rhythms] are discordant, there is suffering, a pathological state (of which *arrhythmia* is generally, at the same time, symptom, cause, and effect). The discordance of rhythms brings previously eurhythmic organisations towards fatal disorder" (25; emphasis added). When rhythms are in disagreement, the flow is interrupted and thus causes disruption and a disconnection—or even breakage. For instance, given its nature, the AT&T crash is an arrhythmia by definition. In reading the corresponding network stories, the embodied rhythms reflect arrhythmia while contextual and applied rhythms polyrhythmically work together to achieve eurhythmia. While said eurhythmia does not annihilate arrhythmia nor does it fix it, a story about the arrhythmic nature of the network crash pathologizes the network to audiences, without much explanation as to the nature of healthy rhythms of networks. The embodied rhythms within the AT&T network crash stories unpack the problematic aspect of the event, highlighting statements and responses that discuss the network in a pathologizing tone, not necessarily as a one-time event, but more as anticipated outcomes and consequences. In an article from the United Press International (UPI) archives, Don Mullen cites the AT&T company explaining that "a logic error in [the] computer software caused the nine-hour long-distance line failure that jammed an estimated 50 million calls and also affected 800 service numbers and computer lines" (n.p.). In this explanation, the author identifies the cause ("logic error in [the] computer software") and the effect ("nine-hour long-distance line failure"), implying that the crash was caused by a disruption of the software's "secret rhythm" (Lefebvre 27) and affected the company's and customers' "public rhythm" (Lefebvre 27).

Lefebvre defines the "secret rhythms [as] [f]irst, physiological rhythms, but also psychological ones" (27). In the case of the AT&T crash, the physiological may be equated to hardware, and the psychological to software. On the other hand, Lefebvre defines "public (therefore social) rhythms [as c]alendars, fêtes, ceremonies and celebrations; or those that one declares and those that one exhibits as *virtuality*, as expression" (27). Although not directly defining the social effect of the AT&T crash, a technological application—or translation—of Lefebvre's public rhythm may be argued to mean the collapse of the company's public service, which represents the *virtual expression* of the company as a body (being its connection to the public and social). Another example of arrhythmia induced in the crash story is that "the problem lingered and was expected to take several days to correct" (Mullen n.p.), which is simultaneously a "symptom, cause, and effect" of the "pathological state" (Lefebvre 25). That is, the discordant network rhythms have caused and resulted in the prolonged problem, which in itself has also aggravated the crash consequences and costs.

Mullen's story also establishes contextual rhythms, by explaining the space–time setup of the event.[19] He specifies the central point of the space within which the event took place. The

---

[19] As quoted above, Mullen retells the details of the crash as follows: "Monday's failure was centered in a signaling system housed in New York City that notifies the network which one of several possible routes a call will travel. The system apparently sent a message to the national network saying that all circuits were busy when they actually were not. The company had suspected a software glitch caused the problem and confirmed its suspicions Tuesday by replicating the logic error in company-developed

fact that the error started at a central point and spread on a national level may suggest a single unified message; however, it still implies a repetitive series of events that allowed for the error to spread as such: the various stations had to process the message and execute specific actions accordingly. The repetition of the message in question is important to highlight because, as per Lefebvre, there is "[n]o rhythm without repetition in time and in space" (16). While the error in itself allows the storytellers to pathologize the network, the story does not necessarily address it as arrhythmia, given that there is no clear description of rhythmic disturbances. Just as "[e]ach segment of the body has its rhythm, [which] are in accord and discord with one another" (Lefebvre 47), the network (considered here to be the body) has a rhythm, as well as segments that have their own individual rhythms. The time of the failure (Monday), in addition to that of the testing (Tuesday) are relevant here, as they allow us to determine that the event is a two-part event, where each part can be isolated as having its own rhythms, though within a space that is common to both and thus to the event as a unit.

Lefebvre further explains that

[a]ll becoming irregular [*dérèglement*] . . . of rhythms produces antagonistic effects. It *throws out of order* and disrupts; it is symptomatic of a disruption that is generally profound, lesional and no longer functional. It can also produce a lacuna, a hole in time, to be filled in by an invention, a creation (44).

The crash of the AT&T international network is an instance of irregularity, which as discussed above has caused various negative symptoms (i.e. effects). The hole in time that was created amongst the AT&T clients needed "to be filled by an invention." In an attempt to remediate and

---

software at AT&T Bell Labs. The tests also determined that the failure had not been caused by a computer 'virus'" (1990: n.p.).

deal with the problem at hand, AT&T claimed to have managed in two opposed ways at two different temporal stages. More specifically, Mullen references AT&T chairperson, Robert Allen, stating that "before the size of the breakdown was determined, operators were instructed not to give out access [phone] numbers for [MCI Communications Corporation] MCI, USA sprint or other long distance competitors. Later, he said, 'We asked (the operators) to offer the numbers to customers in the event they did ask for them'" (n.p.). That is, at the beginning of the crash, the company's decision was to retain their client base, by not allowing them access to competitors. With the aggravation of the problem, AT&T had to resort to redirecting their clients elsewhere. Although this approach was not their first choice, the action the company decided to take affected the rhythm of the event, all the while being affected by it. In a *Los Angeles Times* article, Karen Tumulty explains that Allen "admitted that AT&T operators compounded the problem during the early hours of the system collapse Monday by refusing to give customers the phone numbers they needed to gain access to competitors' long-distance systems" (n.p.). Although this statement is expressing the same problematic decisions, it does so in a way that highlights the decisions' impact on the crash, and on the rhythms thereof.

The growth rhythm of the AT&T network is inherent to the crisis it endured, particularly since "[d]isruptions and crises always have origins in and effects on rhythms: those of institutions, of growth, of the population, of exchanges, of work, therefore those which make or *express* the complexity of present societies" (Lefebvre 53). Given the fact that the technological advances that push networks forward are assumed—or even known—to grow at a considerably fast pace, their rhythms can cause disruption in the spatio-temporal reality that they possess and occupy (Lefebvre 206). Cited in the *LA Times* article, Kleinrock explains that technological progress comes with the price of making people vulnerable (Tumulty n.p.). The vulnerability that manifests in the case of this crash as the interruption of a network's rhythms, also extends to social practices and everyday life by means of the media, in that "the media enter into the everyday; even more: they contribute to producing it" (Lefebvre 57). Put differently, stories—including network stories—situate themselves within the event, to then participate in the reproduction of the everyday around it. Although one can see how the AT&T crash stories contribute to the everyday, a more tangible and current example would be that of Waterfront Toronto.

Ambiguity, as will be discussed below, is at the heart of the discourse around Toronto's smart city. Although the lack of information provided by the project parties is in itself problematic, one should also examine the abstractions and ambiguities within the network stories involved. Lefebvre suggests that media "do not discourse on their influence. They mask their action: the effacement of the immediate and the presence – the difference between presence and the present – to the profit of the latter" (57). When network stories focus on how Waterfront Toronto and Sidewalk Labs are failing to provide the public with enough information, they are turning the focus onto one of their subjects, without necessarily providing their audience with any information that can substitute for the lack of knowledge they are critiquing. The media thus establish their presence in the sphere of the Toronto Smart City project. Moreover, embodied and contextual rhythms come together in application unto the everyday life—i.e. applied rhythms are embodied and contextual rhythms combined. The stories discussing the smart city are mostly interconnected, not only because they cite the same individuals (such as Cavoukian and Doctoroff) but also because many of them have hyperlinks that literally connect them to each other. This interconnection contributes to the presence that stories are asserting in the context of

the smart city, where together they more or less form a unit: in this case, the stories'
polyrhythmia achieves eurhythmia and imposes itself on audiences.

The eurhythmia of contemporary network stories—those of Waterfront Toronto in
particular—is composed of secret rhythms disguised as public rhythms. If media manage to
impose themselves on the everyday, they are training audiences to an automated response to the
topic. Regardless of whether or not the intended response is legitimate, any training of readers is
an issue to be investigated, if not contested. According to Lefebvre, "[t]he informational reveals
only tiny details and results" (64). News articles that are addressing the lack of information
around data privacy and security are not filling the gaps; in fact, they are pointing them out
without giving audiences substantial material on which to fall back. This strategy coincides with
the rhythms of "dressage": "a) The internal activity of control . . . . b) Complete stop; Integral
repose (. . . dead time); [and] c) Diversions and distractions" (Lefebvre 49). First, control in the
case of network stories comes from the power and access they demonstrate, as they argue a lack
of—and demand—publicly available information on data privacy and security. Second, the
integral repose can be either attributed to a small gap in publications around the project as it
evolves (which is not necessarily restricted to network stories), or as dead time within a
particular story. For example, advertisements and "read more" links are integral to online news
platforms. Some publications would use these components as dead time, in that they provide a
relief for readers, so that they can pause or scroll down as they move through the article. Though
this approach may be too subtle, it is still quite effective because it is not only a feature of
"dressage" but also a tool used to keep readers absorbed throughout the piece. Third, diversions
and distractions happen when the focus is solely placed on a particular event (within the major
event, i.e. the resignation of the privacy expert of the smart city project) without really

addressing other issues that could ground audiences, such as definitions, or by placing images (i.e. project blue prints), without much detail or explanation around them. The dominance of the media sector thus comes not only from raising fear in readers by quoting prominent figures—especially Dr. Cavoukian—but also from breaking them in to perceive issues in particular ways. Furthermore, while said references may be enough for people to be concerned—rightfully so—they do raise concerns without effectively offering a space for dialogue.

Although one may argue that response statements that companies tend to issue in the face of scandals or personnel resignation, for example, may be a form of dialogue, they are merely direct responses to specific actions or events, and not to the stories that discuss them. In that sense, "[d]ialogue is reduced to dispute" (Lefebvre 57), which suggests that news articles do not really speak to the different network parties or even allow communication between them. In fact, they participate in a blaming dynamic, where parties simply respond to each other's actions to convince audiences of their "good" standing or intentions. Consequently, this approach puts the news outlets in powerful positions of control: they decide what audiences have immediate access to, whose voice is louder, and even what party is in the right. Nonetheless, "[t]he relationship between forces, which requires the domination of one force . . . is accompanied by a disassembly of times and spaces: of rhythms" (Lefebvre 78). The rhythms of network stories thus extend beyond the story tellers, and involve parties who are invested in projects or referenced in the stories themselves. It is important to note that when network stories become tools to dominate a certain scene, they risk the corruption of the everyday rhythms that are connected to or controlled by them. That is, when such stories are participating as agents of power, they become to audiences what their subjects are: events that strive off of ambiguities and manipulations of the space-time continuum within which they exist. Such stories thus re-present the rhythms of human actions, in ways that are not necessarily representative of the inscriptions that the human actions themselves may leave on people's natural space.

Ambiguities around Networks in the Media

Network stories often employ ambiguity to serve for argumentative and propaganda purposes. Notably, while calling out ambiguities seems to be a step towards raising awareness, it can also serve to shed light on particular aspects in favor of others, possibly allowing for discrepancies in the knowledge circulated. People involved in relevant projects, as well as experts and scholars who study and comment on "smart" projects, often discuss missing information, concealed policies, or ambiguous reports, affecting rhythms of network stories as communicated to audiences.

The AT&T crash news articles available online arguably provide adequate information for the client-base affected in particular, as well as curious audience in general. However, the specificity of the discussion and the particularities of the problem remain more or less concealed. Such articles address the vulnerability of networks, commenting on the competition between the various telecommunications companies, and invoking the politics and power plays in the domain. For example, as mentioned earlier, when reporting on the actual cause behind the crash, Mullen references AT&T in explaining that "a logic error in [the company's] computer software caused the nine-hour long-distance line failure" (n.p.). Although this statement locates the error, it does not provide enough specific information—if any; one may assume that the crash of a network that is "run" by computers will be due to a computer error. The specificity that this statement provides is that the problem was one of software and not hardware. The article, however, does provide additional information, when it points out the geographical location of the issue as it originated from New York City. It also simplifies the error, by explaining that the system that "notifies the network which one of the several possible routes a call will travel . . . apparently sent a message to the national network saying that all circuits were busy when they actually were not" (Mullen n.p.). The language Mullen uses is rather accessible to an audience of various levels of technical literacy, practically mimicking the language around telephone lines (i.e. "busy"). In that sense, less technical details are required in order to provide adequate information to the relevant audience, especially given the timeline of this story—three years prior to Weiser's "Ubiquitous Computing" in 1993.

Mullen explains that in response to a question on who would be responsible if the error was human, the AT&T chairperson Allen responded: "As far as our customers are concerned, I did it" (n.p.). This intentional ambiguity mostly serves as a leadership move to protect AT&T employees, though it could also serve for different political and power-play purposes. In fact, Mullen starts his story by commenting on the fact that this network event "underlined the bare-knuckles fight for customers going between AT&T and its competition" (n.p.). The opening of this article emphasizes the fact that telecommunications companies are in a power struggle among each other, and by consequence, failures feed into the success of one over the other. Peter Krapp argues that "[a]uthoritative power is dispersed as global organizations transition from corporations to networks, and citizens of information society are governed less by concentrated coercion and more by ideological power, as manifest in the symbolic practices and norms of cyberspace" (27). Though written twenty-one years later, Krapp's claim corresponds with Allen's move: as the competition between network communications companies increased, the authoritative role (represented here by Allen's) transitioned into a manifestation of the ideological power that Krapp suggests is governing. Openly stating that the citizens are not to find out whose fault the crash is, puts Allen in a position of power that entitles him to *protect* both employees and citizens from the event's particulars. Moreover, Tumulty provides details

that seem to be rather elaborate, but not specific enough. For example, she explains that "AT&T

said the problem was caused by a 'software bug' in computer equipment that processed signals.

The problem spread to the remaining switches, causing congestion that ultimately spread

throughout the system and forced the system to think it was busy when it wasn't" (n.p.). The last

part of this statement ("forced the system to think it was busy when it wasn't") is the clearest

part, due to its simple language, and possibly lack of technical details. The information provided

in the statement, however, is complex. Although Tumulty defines what a software bug is, she

does not explain other technical terms including "switches" and "congestion." While these terms

may be considered common knowledge today, they most likely were not in 1990.

In addition, providing details of the sort, without providing specifics leaves

technologically literate/savvy audience with vague references. More so, explaining that the cause

is a software bug, without specifying the type (i.e. a feedback loop in this case), and not naming

the "equipment that processed signals," is not enough information to understand the process nor

the system, nor the error for that matter. Moreover, this lack could actually impede the

understanding of audiences and counterproductively complicate the message. Further, Tumulty

references Allen as he explained that "[i]ronically, . . . the software glitch that triggered [the]

breakdown appears to have been located in a new feature that was designed as a backup, adding

redundancy that was supposed to make the entire network reliable" (n.p.). This statement is

ambiguous and vague in several ways. Pointing out the irony in the location of the error assumes

an audience that understands the process. In addition, the description of the feature containing

the error is complex yet too concise, which makes it difficult for readers to understand or

decipher the message. More so, describing a feature as having two jobs that are of fairly great scales—backing up the system as well as "mak[ing] the *entire network* reliable" (Tumulty n.p; my emphasis)—can mystify the problem and suggest yet another form of a magical tale for audiences to believe.

In 2014, **Olenick and Associates, Inc** published an article as part of their "Infamous Software Bugs" series, discussing the errors in the AT&T switches. Given that the publication is dedicated to exploring the errors, readers expect a certain amount of details and (more or less) accessible information. Nicole Gawron opens by explaining that the shutdown was caused by "a very complex software upgrade" that was supposed to provide increased speed and better service. In addressing the error, Gawron explains that a 4ESS switch (named Switch A for the purposes of the article) "[a]t one of AT&T's 114 switching centers, . . . experienced a minor mechanical problem and sent out a congestion signal (basically a 'do not disturb' message) to switches it was linked to" (n.p.). As per both Mullen and Tumulty, the error happened in New York; however, Gawron chooses to keep this information ambiguous. Further, the details of the congestion message remain unknown, though admittedly the lack thereof could serve for a better understanding, especially if the story is targeting the general public. The author continues to elaborate on how and why the update caused the crash, by comparing the before and after processes of the switch.

Although Gawron explains the upgrade by using rather simple language, her explanation is vague, especially since she uses the word "message" in a contradictory way:

In order to increase the speed of the calls, AT&T had tweaked the software to send *messages* faster and more efficiently. Switch A did not have to send a *message* saying it was back in service. The reappearance of traffic from Switch A would tell the others that it was working again, and as the switches receive *messages* from A, they would reset themselves (Gawron n.p.).

The ambiguity in this particular case may be the result of non-technical writing. To simplify the information, using the word "message" with two different meanings makes the explanation less intelligible. The article continues with a more elaborate explanation of the breakdown process, expanding on how the other switches' responses cascaded the effects of the glitch; although it provides details on the process itself, this explanation does not provide any more clarity on the glitch. Gawron ends the article with what seems to be a summative section titled "the reason," in which she states: "[a] single line of buggy code from the complex software upgrade, related to the fact that the switches were receiving messages too soon during the reset process, caused the cascading switches failures" (n.p.). She also offers a link to the pseudo-code containing the bug. It is important to note that the closing of the article arguably provides a more concise and direct explanation of the problem. Although not very thorough, providing a clear explanation leaves readers with a curiosity to pursue more information if they feel so inclined, all the while avoiding unnecessary details that may actually deter audiences from the conversation.

A similar approach is employed in film and TV, where ambiguity can serve for different purposes and have different effects, not all of which are exclusively negative. In discussing stories about Hermes, Alexander Galloway tells that "During Hermes's tedious monologue, all of Argus's many eyes gradually close in sleepiness, and his fate is sealed. Argus, in essence, was bored to death by the most boring thing of all, tales about technology" (36). As per Galloway's example, ambiguity in storytelling can be deemed necessary, if only to keep audiences alert and interested. *Halt and Catch Fire* experiments with fiction and nonfiction in the telling of its stories. While it may seem that the show is providing rigorous details, it actually vaguely replicates non-fictional instances, while focusing more on the fictional everyday lives of the characters. However, concerning the purely technical details, particularly those related to the ARPANet project, the show gives audiences enough to keep them interested without overwhelming them with excessively technical information and distracting them from the storyline. When Ray and MacMillan are working on the network that is the "next big thing," the focus is mostly on the efforts they put in the process. For example, in one of the scenes, Ray is portrayed brainstorming and staring at a board with a web made of yarn. Although it is clear that the project in question is a network (visually and physically represented in this case), the main concern is not the critical details of the network representation, but the intensive intellectual and analytical processes that Ray goes through.

Furthermore, in the scene where MacMillan reaches the epiphany of creating a regional network (S3:E5), the conversation is set around not only innovation, but also finding and filling the gap in people's technological and networking needs. In his brief monologue, MacMillan is speaking more to the "dream" aspect of the project than to that of its reality. When Ray tries to argue, bringing up slight financial and technical details, MacMillan ends the conversation asking Ray if he is hungry, interrupting his rather frantic rhythm associated with the intricacies of the network. Ray's quick reaction reiterates his dedication to MacMillan himself—"I don't even know if you like me!"—and ends his speech saying "I feel crazy!" In response, MacMillan assures him that he is only hungry, and that he himself has "had a tough week" (S3:E5). This dynamic takes audiences back into the story of the characters and their personal lives, keeping the technological content ambiguous. In doing so, the show keeps audiences most engaged with and hooked to the stories of the characters, yet appeals to their hunger for some technological substance deemed trendy, interesting, or even exciting.

Ambiguous technical content is also at the heart of most conversations surrounding the most contemporary project—Toronto Smart City. Given that Sidewalk's plan is to create an urban network environment, the question of data and privacy is more of an obvious concern, yet a tricky subject to make public and discuss openly at a detailed level. **Reporting on parties gaining from Sidewalk Toronto, Jim Balsillie explains that "[f]rom the** start, this project should have been debated publicly and involved experts in [intellectual property] IP and data. Instead, Waterfront Toronto continues to *weaponize ambiguity* while making irreversible decisions that will have major negative effects on all Canadians" (n.p.; my emphasis). Some may argue that this statement is not entirely true, given Waterfront Toronto's and Sidewalk Labs's public and social media presence, in addition to their promised clarity and information.[20] However, neither Waterfront Toronto nor Sidewalk Labs provide the amount of information necessary for the public to understand how their data is being processed. For example, in a tweet published on March 3rd, 2019, Waterfront Toronto links to an information sheet "where [they] are addressing some of [the] questions" around data collection, storage, use, and consent (@WaterfrontTO, tweet). The information sheet restates some of the questions it claims to be addressing (including "who is collecting what information about me? What are they using data about me for? How do I provide or revoke consent for the use of my personal data?") (Quayside Civic Labs 1). However, instead of effectively answering the listed questions, the information sheet explains and defines the following notions: data residency, the "right to be forgotten", algorithms, and consent. Interestingly, these sections provide general definitions, examples outside of the Quayside project, and statements such as "there are important discussions to have about how personal data

———————————————

can be collected, stored, analyzed, and transmitted in smart cities, while ensuring users have the ability to meaningfully consent to participate (or not) in these actions" (Quayside Civic Labs 3). The information sheet thus simply reiterates the concerns and questions that the public has, without answering them or providing any information to which audiences do not already have access.

According to the Sidewalk Labs Project Update published on February 14, 2019, the company promises "a new standard for data privacy including the establishment of an independent Civic Data Trust, which [they] hope will serve as [a] model of urban innovation around the world" (Doctoroff 4). As Doctoroff maps out the company's digital innovations, he expands on the Civic Data Trust, explaining that "[a]ll data collected in the public realm is de-identified of all personal markers, and made publicly accessible through published standards" (4). Although this claim may sound reassuring to its readers, it remains vague and ambiguous to those of them who are not tech experts or enthusiasts. Terms like "public realm," "de-identified," and "personal markers" are not common knowledge and require explicit definitions. In addition, the notion of publicly accessible data does not necessarily convey a sense of trust nor privacy; a more detailed explanation would make the approach more concrete.

In the section on the "new standard for data privacy," the report defines the Civic Data Trust as "[a]n independent entity to control, manage, and make publicly accessible all data that could reasonably be considered a public asset, and a set of rules that would apply to all entities operating in Quayside, including sidewalk labs" (12). First, "independent entity" is a vague term that needs to be more specifically defined, particularly because it could have a variety of different meanings and suggest both hardware and software ambiguities, raising questions that include but are not limited to: what is this entity? Where and how is it stored? who has access to it? who manages it? and what is its structure? Further, controlling, managing, and making publicly accessible is a process that is concerning, especially given how closely it corresponds to network protocols as means for control, as defined by Galloway.[21] The Civic Data trust controls and manages the city/urban environment of the Toronto waterfront, which in itself is a heterogeneous material milieu. The description thereof, thus, contributes to the control effect, as it only answers the "what" question, without addressing the "how" (process and praxis), assuming a pre-existing understanding of the purpose.

The standard for data privacy also promises "responsible data use assessment" (RDUA), described as a "[p]ublicly auditable assessment for all public and private digital services required before data is collected and used" (12). As per other descriptions, the notion of public accessibility—along with the role that the public plays—is highly emphasized, yet seldom defined, if ever. Hence its connotation remains vague, though one can argue that the mere mention of the concept can be enough to induce a sense of trust and comfort in the general public. In addition, the scope of this particular rubric amongst others, is broad and undefined: "all public and private digital services" can mean a wide range of services, and are arguably difficult to discretely and concretely trace. That said, even though Sidewalk Labs promises to "publicly" audit and assess data, it leaves ambiguous the method that will actually guarantee the sharing of adequate information with the public—particularly with people who are not residents of the smart city. Conversely, while the company may actually be treating the data "responsibly," the general public still has little access to the detailed process and to its consequences. In relation to the RDUA, the "responsible data use guidelines" (RDU Guidelines) guarantee an

---

[21] "Protocol is a distributed management system that allows control to exist within a heterogeneous material milieu" (Galloway 2004:8).

"[a]pplication of the guidelines to all parties in Quayside, not just Sidewalk Labs, to put personal privacy and the public good first, while fostering innovation" (12). As reassuring and assertive as this statement appears to be, it is not elaborate nor concrete enough, and therefore remains problematic and unbinding.

Gabrielle Canon quotes a Sidewalk statement addressing the resignation of their privacy expert Cavoukian, in which the company explains that it is committed to its data privacy strategies, though it cannot guarantee the commitment of other companies involved in the Quayside project, because whether or not they "would be required to do so is unlikely to be worked out soon, and may be out of Sidewalk Labs' hands" (n.p.). Although this statement dates back to 2018, it disrupts the credibility of the February 2019 project update, because it is more definitive and consequential. Further, the ambiguity observed in the Sidewalk reports as well as the Waterfront Toronto info sheet is only fueled in media articles and stories, as they contribute to the abstraction of the issues at hand. In an editorial opinion piece by The Guardian, the author claims that

> [s]ensors and cameras everywhere will record the physical world – pollution levels, traffic flow, weather and so on – and of course the behaviour of its human inhabitants . . . . The combination of fixed physical sensors with the mobile ones we carry around with us all the time in the shape of our phones would yield an unimaginable richness of data (n.p.).

The generalization, possible exaggeration, and vagueness of this statement serve to highlight the seriousness of the problem, as well as to push audiences to take a stance. Nonetheless, such an ambiguous, rather fear-inducing statement adheres to the language of the Quayside project companies themselves, leaving the general public with limited understanding and resources.

Further, people are not necessarily provided with sources against which to compare the quality of information they are receiving, and so are restricted to whatever stories the media provide. Therefore, since the project parties and spokespeople are responsible for the ambiguities they generate around the events, demystifying and disambiguating remains the task of the media and network storytellers.

Effects of Networks as per the Media (Real and Hypothesised)

Across the various network stories explained above, the common notion of the effects of technology seems to dominate, even when addressed in various rhythms—at different scales and various levels of seriousness. Further, depending on who is telling what story, the effects addressed may be positive, utopic, negative, or disastrous. The negative and disastrous effects of networks as discussed by the media often pathologize these infrastructures, even when they are not actually pathological. That said, "[w]e are only conscious of most of our rhythms when we begin to suffer from some irregularity" (Lefebvre 86), which explains why most stories are developed around pathologized aspects of networks. The AT&T crash is a prominent example of the negative, if not disastrous effects of networks, even though it was simply caused by a glitch. The stories that reported on the crash emphasize the various losses that were suffered as consequences, including financial, economic, business, and even social.

In his article, as quoted above, Mullen describes how the AT&T error interrupted millions of calls and hundreds of service and computer lines (n.p.). More importantly, the article quotes Allen explaining that "[t]his was the first time in the 32 years since [he had] been in the business that [they have] had this serious a failure" (Mullen n.p.). Placing such an emphasis on the magnitude of the failure, and quoting the chairperson of the company in question only emphasize the seriousness of the consequences and effects. Reporting on the tangible effects of the crash, scholars explore financial as well as socio-economic problems. More specifically, Dennis Burke claims that

> almost 50% of the calls placed through AT&T failed to go through. Until 11:30pm, when the network loads were low enough to allow the system to stabilize, AT&T alone lost more than $60 million in unconnected calls. Still unknown is the amount of business lost by airline reservations systems, hotels, rental car agencies and other business that relied on the telephone network (n.p.).

Providing more or less specific and high numbers, even when simultaneously mentioning the "unknown" amount of money lost by the many businesses, makes rather tangible the cost of a failure/problem so vague and abstract. Similarly, Olenick and Associates, Inc's website refers to the effects as "The Cost:" they estimate a "loss of $60 million in long-distance charges, 9 hours of service time, approximately 75 million missed phone calls, and an estimated loss of 200,000 airline reservations" (Gawron n.p.). Accordingly, the cost is not only assessed in monetary terms, but also in terms of missed phone calls and other forms of transactions (such as reservations), suggesting that the effects of network failures extend beyond financial inconveniences for the big companies, to affect the everyday life of the general public. Further, the socio-economic effects, although quantitative at times, are not necessarily subsequently tangible.

In her article published on January 17, 1990 (two days after the crash), Tumulty refers to people affected by the breakdown as "victims," explaining that AT&T "might make restitution to some" of them, in addition to the possibility of "offer[ing] all customers one day of discounted rates" (Tumulty n.p.). Interestingly, the article claims that the crash in itself raised awareness among the public, about the "invincible" state of "the nation's increasingly sophisticated telecommunications systems" (Tumulty n.p.). As such, it is important—and only realistic—to acknowledge that technologies are not *actually* magical and their systems are prone to errors, failures, and disruptions. Of note, Acknowledgements of the kind, however, are not as common as need be in technology media. Tumulty further references experts who argue that such errors and malfunctions "are inevitable as the nation and world grow more dependent on giant computers to handle everything from banking to national defence" (n.p.). The technological boom at the time of this crash posed a lot of questions and higher expectations than what was actually feasible, contributing to the notion of technology as magic; therefore, corrective statements like these were necessary to update the public on the actual state of technical and networking progress. More specifically, the article quotes Kleinrock: "We're vulnerable . . . . That's not to say this technology should not proceed, but it's not without its price" (Tumulty n.p.). This statement somehow humanizes the technical field—though not necessarily the technologies themselves. Claiming vulnerability creates a connection between people and network companies, asking people to make room for error, or at least allowing them to anticipate failure. Complementing the statement with a disclaimer that technological vulnerability should not impede technological progress replicates a more or less human condition that progress and evolvement do not conventionally come without a price.

Halt and Catch Fire presents an extreme example of the price that people could pay due to technological advancement and how vulnerable it can make them. In the third season of the show, MacMillan, then the CEO of MacMillan Utility, and Ray work together towards "the next thing." After refusing to charge the users for the MacMillan Utility anti-virus software, MacMillan recruits Ray, to work privately with him towards an alternative income stream. After a series of secretive and deceptive behaviour, McMillan Utility board removes Joe McMillan from his executive position, which also affected Ray's work and position. As a result, and in a reaction to the politics of the business, Ray leaks the source code of the anti-virus software, violating the Computer Fraud and Abuse Act (CFAA). This violation puts Ray in a critical situation with law enforcement authorities including the Federal Bureau of Investigation (FBI). After many struggles, Ray commits suicide, leaving a note titled "You Are Not Safe" (S3:E8). The note in itself is a warning against the hypothetical effects that networks will have, especially if people did not act in a preventative way. Ray begins his letter warning that no one is safe, and that "'security' is a myth." Being the "creator" of the network that will allow open and limitless connections, Ray can anticipate the consequences; he is most aware of the price to pay—after all, the character's death is arguably a direct consequence. Warning that "the world is about to crack wide open," he anticipates the loss of boundaries, and the limitless opportunities that are about to become possible. Further, he follows the promise of "a massive connectivity" with a warning against vulnerability, a price to be paid, as well as the impossibility of even "pretend[ing] that we can protect ourselves."

The letter is clearly cautioning against loss of privacy and non-consensual data accessibility, an issue prominent in current conversations and debates around artificial intelligence. Nonetheless, Ray ends his message stating that "[i]t's a huge danger, a gigantic risk, but it's worth it. If only we can learn to take care of each other. Then this awesome destructive new connection won't isolate us. It won't leave us in the end so… totally alone" (S3:E8). This final promise is a simultaneous warning that people need to anticipate this change and address the technological advances in unity, against what could otherwise be disastrous and destructive. The closure of the letter demands that people shield themselves from the isolation that the new communication networks may cause, by actually protecting each other. The promise, however, is a positive outlook on the potential of technology, and a hopeful message regarding the power and possibilities of networks. The *Halt and Catch Fire* example demonstrates some parallels to the less fictional and more contemporary example of Sidewalk Toronto.

The effects of the Toronto Smart City story became less hypothetical as the project came closer to its presumed realization. The tone around Sidewalk Toronto shifts from what was positive and optimistic (with the announcement of the project in 2016), to what later became threatening and pessimistic as the project evolved. In April 2016, *Business Insider* published an article titled: "Google Wants to Build its Own Futuristic, Smart City." Danielle Muoio opens the piece by explaining that Sidewalk Labs's "aim is to transform parts of economically struggling cities into a proving ground for cities of the future" (n.p.). This statement builds solely on information available on the S*idewalk Toronto* website, whose "Objectives" section promises:

- "Establish[ing] a complete community that improves quality of life for a *diverse* population of residents, workers, and visitors.

- Creat[ing] a destination for people, companies, startups, and *local* organizations to advance solutions to the challenges facing cities, such as energy use, housing affordability, and transportation.

- Make Toronto the *global hub* of a rising new industry: *urban innovation.*

- Serve as a model for *sustainable* neighborhoods throughout Toronto and cities around the world" (Sidewalk Toronto; my emphasis).

The above listed objectives emphasize community and innovation: Sidewalk Labs advertises using technological innovation to enhance community and urban life as their major goal. Further, the objectives employ keywords such as "diverse," "local," "global hub," "urban innovation," and "sustainable," to appeal to geographical and contemporary concerns. The main target audience of Sidewalk Toronto would be Torontonians in general, and more specifically the hi-tech enthusiasts among them. Therefore, addressing issues that are of major concerns is the way to appeal to the particular demographic in mind. However, in February 2019, a quick Google search of the expression "Toronto smart city" shows the first non-ad result to be an article in *The Guardian*, titled "'City of Surveillance': Privacy Expert Quits Toronto's Smart-city Project," with a subtitle that reveals effects to be addressed: "Wired neighborhood planned by Google sister company has raised questions over data protection." Published on October 23rd 2018, the article tackles the resignation of Cavoukian. Canon begins her article with a reference to critics' concerns upon the announcement of the project within the context of new, smart, high-tech models. The following excerpt is one that frames Justin Trudeau's comments in an opposition statement:

> Despite Justin Trudeau's exclamation that, through a partnership with Google's sister company Sidewalk Labs, the waterfront neighborhood could help turn the area into a "thriving hub for innovation", questions immediately arose over how the new wired town would collect and protect data (Canon n.p.).

The juxtaposition of a more or less romantic language ("*thriving* hub for *innovation*") to what seems to be a realistic concern ("questions" about data collection and data privacy), sets the tone for a justified worrisome text to follow, especially given the news it is reporting. Canon reports that Cavoukian resigned in order to "'send a strong statement' about the data privacy issues the project still faces" (Canon n.p.). Not only does the resignation news call for serious concerns, but the statement that accompanies it legitimizes skepticism and alarms around the smart city, implying a continuous failure in addressing the issues that had been raised even at the very early stages of the project. Adding to the magnitude of the problem at hand, Canon quotes

Cavoukian's resignation letter, in which she opposes her vision of the smart city as one of privacy to what is actually the current status: "a Smart City of Surveillance" (Canon n.p.).

Surveillance occupies an important part of the conversation around Toronto Smart City. In his article, Balsillie claims that Sidewalk's "business model is built exclusively on the principle of mass surveillance" (n.p.). He continues to state that the growth of Alphabet—Google and Sidewalk's parent company—and its future rely heavily on intellectual property and data (which it controls) (Balsillie n.p.). This statement suggests that the company *needs* to own the data and IP, which is problematic as people's information becomes the property of a big company that can sell it, invest in it, and make it available to other parties—that is in addition to the obvious concern of privacy invasion. Balsillie also addresses the fact that Google "tracks users even when they [do not] consent to being tracked" (n.p.). Ignoring whether or not the user consents to having their information recorded is a violation of the user's rights, bordering information theft.

As Balsillie puts it, Sidewalk Toronto "is a colonizing experiment in surveillance capitalism attempting to bulldoze important urban, civic and political issues" (Balsillie n.p.). Through its strong and somewhat violent rhythms, this statement serves as a warning against effects that are not only political in nature, but also a threat to people's everyday lives (with respect to the urban and the civic contexts). In light of the media attention that the project attracted, Doctoroff responded with a blog post—on February 14th 2019—stating that "[r]ecently, some of what [they have] been working on has made its way into the media and so [they] wanted to share directly with [audiences] some of these key concepts as they are coming into focus" (Doctoroff n.p.). Although this statement promises audiences content that is relevant to what is being circulated in the media, the blog does not really address any of the privacy and data issues that are actually of concern to people. Further, the post repeatedly assumes excitement on behalf of Torontonians, which may present as a statement that disregards opinions and concerns that media outlets and people are openly discussing. Doctoroff's post, although repetitive, does not really assume a rhythm: "[t]he monotonous return of the same, self-identical, noise no more forms a rhythm than does some moving object on its trajectory" (Lefebvre 86). This non-rhythmic—and rather noisy—attitude is the consequence of what Jodi Dean calls "Communicative Capitalism" (2005), explaining that "[u]nder conditions of the intensive and extensive proliferation of media, messages are more likely to get lost as mere contributions to the circulation of content" (Dean 53-4). In other words, when Doctoroff repeatedly emphasises how he "thinks" that Torontonians will be as excited as his team is, particularly in a post that is supposed to address media and people's concerns, he is overriding the information already out as a message only to treat it as a "contribution" (Dean 58), thus "merging . . . democracy and capitalism" (55). Instead of addressing the content of the messages—privacy and data concerns—he is addressing the contribution, as if the media is merely discussing Sidewalk Toronto, and that people are learning about it, without acknowledging the actual fear-inducing content of the stories.

Conclusion

When the media presents network stories to the public, they often focus on the effects that network events have on societies and people. To intimidate and dominate, media outlets impose their power and control via secrecy and ambiguity, granted that their audience may not have the expertise to understand the complex notions that they may provide. Network stories thus do not help their audiences to participate in the events on which they are being informed. In fact, the public receives more or less corrupt information, in that it does not provide enough knowledge, yet instills concerns and conceivably unfounded fears. Further, the stories are usually distant from their corresponding events, especially when they only focus on qualitative information, without looking into quantitative data, or merely presenting them out of context. It is particularly important to provide quantitative analysis for the purpose of adequate depictions of the time-space continuum, given the tangible specifics this approach can provide. Further, the relationship between the time-space continuum and rhythms, as determined by Lefebvre, is crucial and thus essential in network stories. The next chapter discusses what network communications sound like, to provide an understanding of the rhythmic nature of networks, offering more insight into their structures. While network stories may linguistically represent the rhythms imposed by corresponding human actions, sonifying networks makes their rhythms audible, as they are carried out by the technologies involved.

Chapter 3: Listening for Network Events

Jonathan Sterne claims that "[t]o think sonically is to think conjuncturally about sound and culture" (3). Listening to sounds often invites cultural dimensions into our understanding of the various layers and dimensions of the environment we live in. Whether one listens to songs of birds on a clear morning, the hissing of tree leaves on a breezy day, the rhythms of waves on a sunny afternoon, or even the humming of machines on a busy day at the office, the soundscape offers different types of information about the space within which we happen to be at a given time. Although people do not necessarily always listen with intention, the three modes of listening as defined by Michel Chion help determine the outcomes of a listening exercise. Causal listening, for instance, "consists of listening to a sound in order to gather information about its cause (or source)" (Chion 25). The sound itself in this mode is a representation of an event, and the intent of listening is identifying the cause behind what we hear as opposed to mere sonic identification. Another mode is semantic listening, "which refers to a code or a language to interpret a message" (Chion 28). In this mode, the sound acts as a message, and listening carries the intention of deciphering and making meaning of the sound we hear. This particular mode is more relevant to linguistics; within the context of this work, semantic listening manifests as a gap between what people know about network communications and the possible information carried in these exchanges. To arrive at a semantic listening to network communications, there would need to be some sort of alphabetic/symbolic or linguistic exchange associated with the sounds of networks, which is not the case and thus makes semantic listening extremely difficult if not impossible. On the contrary, in reference to Pierre Schafer, Chion explains that reduced listening "focuses on the traits of the sound itself, independent of its cause and of its meaning" (29). Reduced listening focuses on the sound as artifact, in an attempt to recognize its pure characteristics outside its potential causes; it allows the listener to pay close attention to the details of the sound in question, repeatedly, until the multiple aspects and events that characterize it are registered, and appropriately recognized and described. In its first stages, this project utilizes reduced listening, followed by an attempt at causal listening once the sounds are appropriately described as independent subjects of inquiry. Semantic listening however cannot be applied to network communications because they do not use symbolic language; it will nonetheless be occasionally addressed throughout this chapter.

In the age of networks, technological devices communicate ubiquitously, leaving minimal tangible traces that people can perceive. These devices emit signals and messages; along with information, they communicate data that serves for their mere functioning—i.e. laptops and Wi-Fi routers are not constantly communicating "meaningful" information of interest to people, but it is necessary data for a device to access the internet. Although the EM waves that materially make these network communications possible are inaudible and invisible to people, they do exist within our environment. It is difficult to question the material presence of network communications, as well as their possible implications, especially given their ubiquitous status. Making concrete the physical presence of EM waves solidifies the fact that when technologies communicate, they do actually leave a trace, inducing more "modification of [the natural] rhythms . . . by means of human actions" (Lefebvre 117). Moreover, sonifying network communications materializes how human technological actions contribute to, if not interfere with, the natural soundscape, re-marking natural space.

Networks rely on algorithms that dictate the events of emitting, receiving, and processing EM waves amongst devices, to make their communications possible. Shintaro Miyazaki explains that we are surrounded by infospheres consisting of vast electromagnetic (EM) networks

created by assemblages of antennas, satellites, cables and other bits of communication

technology for data transmission intermingled with computational devices of data

processing and storage such as smartphones, laptops, netbooks or tablet computers (514).

The world we live in has come to accommodate "infospheres" that are essential for

communications not only between people (in virtual ways), but more deliberately between

technologies themselves, even outside the direct human needs for them: at extended times,

technologies communicate with each other, without a *direct* purpose of serving for human needs

for communications or everyday operations. These infospheres are infrastructures that physically

surround us, and involve people in the techno-environment simply by sharing the same space.

However, the inaudible nature of eventful EM waves makes it difficult for people to

acknowledge their existence, let alone try to understand them (semantically for instance), or

question their effects and consequences. To that end, "[d]ata sonification [presents as] a method

of exploring processes in spacetime terrains such as bodies and landscapes" (Palmer and Jones

222). Sonifying networks can situate their communications and EM waves exchange within a

tangible spatio-temporal experience. Being able to listen to the sounds that the technological

devices produce, provides a new approach to recognizing what rhythms network events

contribute to the space-time continuum of our everyday life. Sterne explains that "sonic

imaginations rework culture through the development of new narratives, new histories, new

technologies, and new alternatives" (6). Using sound as a method to tell new stories about

historical and technological events allows a new understanding of the cultures around these

systems. (Re)visiting the sounds that technologies produce in the silence of their communications, promotes a new discourse around the infrastructures of networks—and the performances of algorithms—within the existing cultural manifestations of power and control.

Michaela Palmer and Owain Jones explain that "sonification artefacts or events retain temporal and performative dynamics within themselves as they play in time" (222). Sounds, as heard and listened to in time and space, interact with the continuum in a way that is performative, through the rhythms that they carry and implicate on any given soundscape. Sonic artefacts in and of themselves have features that represent the nature of networks, particularly in their spatio-temporal performances that represent an integral part of algorithms. Miyazaki explains that "[w]hen an algorithm is executed, processes of transformation, and of transduction from the mathematical realm into physical reality, are involved" (136). The algorithm is processed in a way that reads and calculates mathematical commands, and executes them into the physical reality of the software and—more importantly in the context of this dissertation— network that it defines, through EM waves or hardware, respectively. This feature of algorithms—mathematical translations into EM waves—highlights their performative nature, in that they participate in a process that gets applied into, and contributes to the physical reality of networks. Listening to such processes makes tangible the physical execution of algorithms and the physical nature of networks, permitting a better understanding of how networks and their algorithms work and rebrand the rhythms of our societies. Similar to stories about networks (network stories explored in chapter 2), listening to networks and their communications is another form of representing the rhythms that people inscribe in their space (Lefebvre 1991), reconfiguring the natural space within which they live.

Listening Methods (Modes)

Reduced Listening

I use reduced listening to describe the sounds in a way that is independent from the settings that surround and lead to the production of the sonic outcome. Using raw/literal language to describe sonic events could be helpful in—potentially—identifying what is "meaningful" and what is "noise", if only through what is aesthetically described as positive versus negative. However, R. Murray Schafer explains that "[n]oise pollution results when man does not listen carefully. Noises are the sounds we have learned to ignore" (Sterne 95). This definition may suggest that by exercising reduced listening on transduced EM waves of network communications, one cannot describe sounds as noise, given that we do not normally hear them, let alone have been trained to ignore them. Therefore, reduced listening does not by itself engage in identifying/classifying sound from noise, since its main purpose is to describe the sounds we hear regardless of their meaning or cause. However, reduced listening potentially helps in distinguishing sound from noise in the causal listening stage.

The below analysis does not assume that aesthetics are enough to separate sound from noise, and is open to outcomes that may not necessarily conform to expectations. The questions to keep in mind as one practices reduced listening are: what is at stake when the sound is completely separated from its cause and meaning? What does it mean to listen with the pure intention of describing the sonic object and event? What are the advantages of acousmatic listening in the context of networks? The descriptions below attempt at answering these questions, taking into consideration the possibility that a more accurate understanding may unfold in the next sections, and further research—and projects—may be necessary. To arrive at the most detailed description possible, within the constraints of this project, I listen to clips multiple times, with careful pauses as necessary; repeated listening allows me to hear particular details and events that may have gone unnoticed in the first few times, as I get acquainted with the sound clip being studied.

Causal Listening.

I use causal listening in an attempt to understand the causes behind the particular sonic events and/or noises. I do not claim that I have exact and accurate determinations of what particular technological event or algorhythmic command causes a specific sound. However, I suggest that practice in identifying causes behind particular sounds can lead to a better understanding of the technologies investigated, and to making networks more accessible to the general public. More specifically, I focus on causes behind the rhythms that we hear, as well as interesting sonic event points—i.e. brief sonic moments. Chion urges that:

> We must take care not to overestimate the accuracy and potential of causal listening, its
> capacity to furnish sure, precise data solely on the basis of analyzing sound. In reality,
> causal listening is not only the most common but also the most easily influenced and
> deceptive mode of listening" (26).

Given the nature of causal listening, people may be inclined to assume particular causes to be associated with a given rhythm or sonic event, which in turn leads to deception; this phenomenon is mostly due to the fact that the sounds we hear can be affected by the environmental soundscape within which they exist.

Although I use causal listening to better understand technological processes, I do not argue that this practice is enough to truly comprehend the material investigated. In the context of the listening exercises below, causal listening is guided by the descriptions provided by the artists, though I also introduce some observations of my own. I start by presenting what the artist explains about the particular sound clip—if any information is available—and continue to conduct my own causal analysis. I focus on event points and event patterns, and try to understand what they could mean. The reduced listening descriptions are referenced when attempting to interpret said event points and patterns (during the causal listening descriptions).

Semantic Listening

In its original methodological planning, this project imagined the use of semantic listening to be at the heart of the Tactical Network Sonification technique. Chion explains that semantic listening "is purely differential" (28). Within the context of linguistics, Chion notes that "semantic listening often ignores considerable differences in pronunciation . . . if they are not *pertinent* differences in the language in question" (28). Consequently, having a basis for differential analysis is particularly important in semantic listening. For example, it would be ideal to have a basic semantic knowledge of sounds within network communications, to be able to learn which sounds are noise and which ones are meaningful. Note that the difference between causal listening and semantic listening is that the former identifies the cause—action, element, command—that resulted in a particular sound, while the latter would identify what the sound itself means—for example, (hypothetically) that a given sound means that one machine is sending location data to the network receiver. Before giving up on semantic listening, I tried to explore ways of acquiring a set of sounds that are important within networks. To do so, it appeared important to have recordings of networks as they sound individually: every network has its own sounds, which in an infrastructure may be muffled or overridden. It was technically—and theoretically—impossible to separate networks within a system into separate entities.

Additionally, given the nature of contemporary networks, the soundscape of separate networks would be different than that of the same network within the whole system. Due to the fact that networks operate and communicate around each other, the EM emissions accommodate and account for all the active networks with a system. As such, listening to networks separately and outside of their infrastructure would practically defeat the purpose, because it changes the conditions and circumstances that produce particular algorhythmics. That said, networks are crowded and multi-layered, which in turn means that I could not reliably identify a set of sounds as a basis for semantic listening. This project thus acknowledges that practicing semantic listening of networks, is not feasible, even if listeners were to try and find meaning beyond the linguistic codes and symbols.

Listening Exercises of Online Clips

This section provides a series of listening exercises, using openly available recordings of technological devices. During these exercises, I describe and analyze the sound clips using reduced and causal listening. The analysis does not use semantic listening because decoding these sounds would be making far-fetched assumptions about communications that do not happen within a symbolic or linguistic system. Using semantic listening would also result in claiming that the sounds one hears are not only the product of technical processes and events occurring when technologies communicate with each other, but also a means for technologies to communicate with people via the disseminated transduced EM waves. This project does not claim that technologies can (intelligibly) communicate with people; in fact, the analysis below is an example of how eavesdropping on network communications may allow us to breakdown a given process, though people still need different devices and more tactical methods to arrive at detailed understandings of the network and the events it produces.

The Detektor and the Elektrosluch are two devices that allow people to listen to EM waves, by transducing them to humanly audible decibel levels. In this section, I first use reduced listening to describe a selection of the sound clips recorded by various artists, to arrive at a detailing of what the sonic representation sounds like. I then apply causal listening, to investigate the causes behind particular sonic events and rhythms; this part, as previously mentioned, may be guided by the artists' descriptions, in addition to my analysis of the sounds and their settings. The sound clips I use in this project are open source; they are chosen because of their relevance to the subject of this project: technological devices in action, communicating with a given network. One problematic issue to note is that these sound clips are more focused on individual machines and the sounds they produce (via EM waves), while listening to networks would mostly require listening to recordings of ambient EM waves emitted by multiple machines during their communications process. Although the causes may be different, reduced—as well as causal—listening remains a similar exercise between the two types. Learning to identify and describe sounds outside of their causes is independent of the frequency range the sounds fall in. Similarly, the process of identifying causes behind specific sound events and noises is not quite different between individual devices and network communications; that said, the causes may be clearly different. Both individual EM emissions and network emissions have interferences that may be external to the process in which the listening exercise is particularly interested. In this section, I present the reduced listening followed by causal listening of every sound clip.

"Torrent-transmission" (1 minute 43 seconds – Listened to on SoundCloud)[22]

Reduced Listening: The sound clip begins with a steady buzz then increases in pitch to become a scrambly and somewhat interrupted track, where the interruptions and cuts are not completely silent. At seconds three and four, there is an anticipatory sound that escalates into the louder pitch of the clip. When the louder sounds quiet down, the track becomes more staticky and steady with minor rhythmic sounds that are similar in their characteristics to the quieter buzz of the beginning of the clip. The sounds then go back to being loud and somewhat crowded, but they quiet down again for just under a second to then continue with the loudness, still with some minor interruptions. At second twenty-three the sound thickens but with minute squeaks that can be more or less heard in the background of the more dominant sounds; the clip gets un-rhythmic between seconds twenty-five and thirty before it quiets down at second thirty-one. Further, the sound clip quiets down between 1'10" and 1'11", for approximately a full second and then goes back to the—almost—same sounds. It gets more rhythmic, however, towards the end of the clip before it quiets down again. From the minor down times, it sounds like the lower staticky buzz is an underlying constant throughout the clip. The arhythmicality of the sound is dominant, except for some stretches—second twenty-one to twenty-two—of loud uninterruptedness. There is a certain steadiness to the quieter static noise, though one can hear some pulsation. Between minute one and 1'10", the sound gets somewhat patchy and even less consistent with the rest of the clip, leading to silence, followed by a continuation of the inconsistencies until the last static quietness before the end of the clip. Note that, between seconds thirty-six and thirty-nine, there is a layer of static sound that is different from the rest of the sound clip's staticky sounds, mostly in pitch.

____

Causal Listening: This sound clip does not have descriptions or comments; therefore, I base the causal listening exercise on my personal interpretation of the sonic events and their reduced description, supported by necessary research around the process involved. From the title of this clip, one can deduce that data transfer is involved in the process. As defined by TechTerms

> a torrent is a file sent via the BitTorrent protocol . . . . During the transmission, the file is incomplete and therefore is referred to as a torrent . . . . Torrents are different from regular downloads in that they are usually downloaded from more than one server at a time. The BitTorrent protocol uses multiple computers to transfer a single file, thereby reducing the bandwidth required by each server. When a torrent download is started, the BitTorrent system locates multiple computers with the file and downloads different parts of the file from each computer. Likewise, when sending a torrent, the server may send the file to multiple computers before it reaches the recipient (Christensson n.p.).

Given that a torrent download requires the system to locate the computers from which to download the file, the steady buzz can be interpreted as the extended event of buffering or searching processes. Once the source(s) is (are) located, the high-pitched interrupted track would signal the process of data transmission. The anticipatory sound at second three to four can thus be interpreted as the event of connecting to the source computer. Since a torrent download happens via multiple computers, it is reasonable to suggest that the clip's staticky and steady sounds—rhythmic and similar to those of the clip's beginning—could signal the switching of connection from one source computer to another. These patterns are consequently repeated throughout the sound clip. The reduced listening explains that the lower static buzz is an underlying constant, which could suggest that the BitTorrent process is constantly looking for the next connection to establish until the entire download process is completed. The static sound between seconds thirty-six and thirty-nine does not adhere to the torrent transmission process, nor does it conform to the patterns of the sound clip; it is thus interpreted as interference and noise.

"HF-pc lookingforwifi" (16 seconds – Listened to on SoundCloud and Audacity)[23]

Reduced Listening: This sound clip is rhythmic, yet continuous and uninterrupted (with no silences), until second five when it acquires a rhythmic beat and is repeated very similarly at the turn of second twelve. The starting sonic event repeats between the first set of beats and the next—i.e. between seconds seven and eleven. The continuous rhythmic sound resembles a push and pull buzz, while the beats at their original playback speed sound like an insect flapping its wings. The two different sets of sounds are not sonically separate; in fact, the continuous track remains in the background of the beat-like sound. When listening in the sound software Audacity, I slowed down the playback speed from 1.0x to 0.25x, and so was able to count five beats in the first subsets of the beat sets, and 8 in the second; the first subsets are identical in both sets, and so are the second subsets. At a 0.5x speed, the clip sounds like a constant wind blowing, as well as the sound of a helicopter, though not as consistent. Simply put, the beats that constitute the helicopter sound speed up and slow down. At 0.01x, the clip mostly sounds like a heartbeat, though irregular.

Causal Listening: This sound clip has one comment at second five: "looking for new wifi networks." Given that the reduced listening exercise explains how this clip consists of a rhythm that repeats twice, the comment provided is misleading for two reasons: first, it is the same as the title, with two added words ("new" and "networks"); second, it suggests that the process of looking for new Wi-Fi networks is just beginning, though the repetition of the pattern before the comment—and after—suggests otherwise. The repetition of the two patterns suggests that the extended event of looking for Wi-Fi produces the same patterns until a network is found. Based on the information made available, I will assume that the repetitive patterns are all part of the search process—that is, for new Wi-Fi networks. I will also assume that the flapping pattern signals that the Wi-Fi adapter—hardware device installed in computers/laptops—is sending out waves in search of a connection. The continuous buzz may be the hardware waiting for a response. I anticipate a change in sounds and rhythms once a connection is (being) established.

"0016_HF-Backgroundnoise" (38 seconds – listened to on Audacity)[24]

Reduced Listening: This sound clip is staticky and rhythmic until the rhythm fades slowly to be overcome by a high pitch whistle, followed by sort of an increase in volume. A high pitch buzz dominates the sound clip. There seems to be an underlying beat that is too low to stand out on its own. At a 0.25x playback speed, one can hear a quick rhythm—a "trrrrr" sound, the 'r's are rolled in this sonic description—that slows down and speeds up. It sounds like it flattens towards the middle of the sound clip and goes back to how it was, though I cannot be sure of the extent of similarity between the initial and final states. Although this sound clip does not convey an idea that there are many events taking place, it is quite layered and diverse. It is worth mentioning that the overlap between the various sounds that I hear (or not), makes it difficult to describe the clip in a general way. Accordingly, to be able to describe the sound events that are taking place in this sound clip, it is important to listen to it multiple times (over 7 times), focusing on each layer of sound separately. To focus on each layer, it is necessary to slow the playback speed to 0.25x and lower. The first sound I focus on is the overlaying rhythmic beat that expands throughout the sound clip. The first dip in the pace and volume of the beat happens at 0.525s; the beat continues arhythmically with numerous dips, though it becomes more consistent from approximately second five until mid-second twelve, when it continues at a lower volume. The volume of the beat increases again around second 16 and remains more or less consistent. However, upon a more focused listening, one hears that the intensity of the beats changes rather frequently, even if the changes are not very obvious at a first listen. It is worth mentioning that the background and the layered sounds make it difficult for one to clearly identify what the beats specifically sound like, or how consistent they are on their own (as a separate and distinct layer). The underlying layer is a buzz that starts low (barely audible) and increases in frequency. At one point, it gets as high in frequency as a whistle. The background "blow" alternates between a low buzz and a whistle throughout the sound clip. Another layer sounds like strong wind blows that are intermittent and different in intensity. Overall, the layers work together in a way that may suggest that the blowing actually affects the beats—their speed, as well as their intensity and pitch.

Causal Listening: The reduced listening description corresponding to this sound clip is rich and detailed, particularly because of the various rhythms and event points that are present. Yet, the title does not provide much detail about the settings of the recording, and there are no comments that would have complemented the sounds that I hear and describe. The information that I do have, however, suggest that this clip reproduces the background noise of a certain, unclear, process. Further, the description of the clip explains that clip number sixteen (among others) was recorded "using [an] electromagnetic coil" (sonic archeology n.p.), which means that it corresponds to a close-up recording of the machine at hand. Because I do not know exactly what transduced algorhythmics events were recorded, the causal analysis can only be general and purely based on assumptions.[25] Given the multitude of layers and variety of patterns, I suggest that although the title claims that it is background noise, various processes are being heard through the different sounds; therefore, calling it "noise" is likely to be inaccurate and misleading.

"GSM-Incoming-call" (32 seconds – listened to on SoundCloud)[26]

Reduced Listening: This sound clip starts with a rhythm of one low buzz then switches into a low beat and a higher beat that are close to each other in time; this sonic event lasts for about two seconds, over-layering a beep that is present yet difficult to hear. At second three, I hear a louder short buzz, followed by a change of rhythm consisting of a faster pace. With this sonic transition, one can hear multiple layers of sonic events: one flat and constant buzz, a fast rhythm, and intermittent (very low) scratchy noises. The new rhythm lasts until second thirteen. It then speeds up even more, to the extent that the beats somewhat sound like a continuous tone; the sound gets a little higher in pitch. At second twenty-one the volume gets noticeably louder, though maintaining the same rhythm. At second twenty-seven the clip quiets down remarkably, with intermittent quick buzzes and scratches, as well as a constant hardly audible whistle. Since this file is not downloadable, I could not listen to it nor manipulate its speed using a sound software (for copyright reasons, as well as technical limitations).

---

[25] Note that this recording was made nine years ago (at the time of writing), so a replication of it will not be accurate given the technological changes that have happened during that time.

Causal Listening: The title of this sound clip is clear, which is somewhat helpful in describing and understanding what is causing the sonic events and rhythms to which I am listening. The quiet buzz with which this clip starts may either be irrelevant given how brief it is, or can signal the period prior to an event—i.e. the incoming call. The two-seconds-long pattern consisting of the low and high beat may signal the beginning of the connection or reception of the call. Between second three and second thirteen, what I hear may be representing the ringing—the connection is established and the caller is waiting for the person on the other line to pick up. At second thirteen, the rhythmic change indicates a status change, which is probably the acceptance of the call. A phone call entails an exchange of data, which explains the layers of louder flat buzzing, fast rhythms, and intermittent low scratches. The increase in volume at second twenty-one may suggest either a solid establishment of connection, a 'glitch' with the listening device that caused it to change the output volume, or even a change in the recording settings, which can also lead to the same outcome. When the clip quiets down at second twenty-seven, the call is likely coming to an end, and the flow of data is thus being terminated. The remaining buzzes and scratches can indicate various elements such as delay in disconnection, or simply interference, noise, or any other—probably unrelated—technological process captured by the transducer.

Case study using the Detektor

Using Miyazaki and Martin Howse's Detektor, a Focusrite Scarlette 2i2 audio interface, and the Sonic Visualiser software (Cannam, 2005-2019), I recorded five, approximately fifteen-minute-long sound clips at the Humanities Computing and Media Centre (HCMC) in the McPherson Library building at the University of Victoria. I have received consent from the centre personnel, as well as ethics approval, to conduct and share these recordings. To maintain a semi-controlled environment, I recorded on Tuesdays and Thursdays, from 1:00 pm to 1:15 pm, while situated in the same cubicle. The HCMC is an open-space computer lab, with twelve computer stations in its main area (ten Dell computers with UNIX operating system, and one Apple computer with an Apple operating system—iOS). These stations are occupied by a fluctuating number of research assistants and project primary investigators. The main area also has a table at its centre, where individuals work on their own laptops (that run on various operating systems). In another area, there are two Dell computers and two Apple computers, with Unix and Apple operating systems, respectively. The area in which I was situated at the time of the recordings has one Apple computer with an Apple operating system, in addition to a number of screens that could be connected to personal laptops via VGA or HDMI chords. The lab operates within the premises of the university library, and is therefore within the connection range of its Wi-Fi. Given the diversity of the HCMC environment, I was only able to record the number of people present at the time of each recording, allowing for the possibility of a person coming in or out (without being noticed or recorded as present). Further, I noted my own personal devices that I have running, in proximity to the Detektor, at the time of the recordings. For the purposes of this section, and given the lack of material for causal listening (as will be addressed below), I will present reduced listening descriptions of each session, and will then discuss causal listening collectively. Semantic listening remains completely outside of the scope of this section, given the complexities of the sounds recorded, and the—purely—assumptive narrative that such an analysis would risk attributing to the communications processes recorded and discussed.

Reduced Listening

HCMC Session 1[27]

During this session there were seven people in the space. The devices running in close proximity to the Detektor were: a laptop, a screen, a Bluetooth mouse, Bluetooth headphones, and a cellphone. Listening to this session, I have identified thirty events; I assume that some similar events have gone unnoticed or undiscernible under the various competing sound layers, which you might be able to identify as you listen carefully. Some of the noted events are patterns or underlying layers of sound that are dominant in the session. Given the repetitive nature of sounds and events recorded, I refrain from describing them individually, though I attempt to describe what they sound like and approximate the number of times these events occur.

There are approximately seven types of events, most of which occur multiple times, at different volumes and intensities; you might find that some events simultaneously embed other types of events within them. The first type of events is a buzz that sounds like the cellphone-interference we hear through external speakers (when our phone rings). This event happens around fifteen times, at different volumes, intensities, and lengths. The second type of events is an intense buzz; it occurs three times in total. The third event recorded is one that is more or less a dominant pattern in the session, consisting of a rather monotonous buzz that fluctuates in volume. The fourth type involves events that sound like scratchy interruptions of more consistent layers. This type occurs approximately three times. The fifth type of events is one that embeds a number of rhythmic occurrences: one-second-long minimal events that happen almost every second; though not very regular in terms of collective rhythms, their individual rhythms are more or less maintained. This type reoccurs twice. The sixth type of events is one of a scratch—low frequency—that is approximately one second long. This event reoccurs once. The seventh type of events is a very subtle change in the beeping, or constant whistle, and is a one-time occurrence. The beeping, or constant whistle, as you hear throughout the clip, presents as a predominant sound, along with a hum. These two sounds are mostly consistent, though they are—at times—interrupted by other dominant sounds and other not so dominant events.

HCMC Session 2[28]

During this session there were eight people present in the room. The devices running in close proximity to the Detektor were: a laptop, a screen, a Bluetooth mouse, Bluetooth headphones, and a cellphone. Music was streaming onto the headphones. In this session, similar sound events as those of session one are detected. When you listen to the track, you may notice that the underlying whistle is almost the same. The hum—also noted in session one—becomes audible to a certain extent, when other sounds quiet down, but is not as prominent as the hum heard during the first session. Further, there are intermittent high-pitched beeps, as well as subtle changes in rhythms more consistently noticeable in session two than they are in session one. Nonetheless, there are three notable events you will hear during this session. The first event is rhythmic, with three extended buzzes, somewhat beat-like. The second event is a buzz that is similar to the first type of events detected in session one (a buzz that sounds like the cellphone-interference); this event occurs numerous times throughout the session. The third event is one that is similar to the second sonic event of this session, though longer; this buzz is also muffled

———————————————

by more dominant sounds. Listening closely to this session, you will notice that it has a number of minor events that prove difficult to distinguish and individually group or even detect.

HCMC Session 3[29]

During this session there were six people in the space. The devices running in close proximity to the Detektor were: a laptop, a screen, a Bluetooth mouse, Bluetooth headphones, and my phone. You will notice that this session has very steady rhythms and sounds, but is also rich in subtle and minor events. As you listen, pay attention to a number of event types noted here, and notice how some of them occur at multiple instances. The first event consists of three beeps—more accurately described as high-pitched whistles—of approximately one second length each. The second event is a brief scratchy buzz that is rather common throughout the session—and is also present throughout other sessions. The third type of events is an intensified buzz, that is higher in volume than others previously detected. This event reoccurs approximately three times. The fourth type of events is an intensified whistle that becomes more like a high-pitched beep. As you will hear, this event is notably repeated at least three times. The fifth type of events is one that is similar to the first type from session one (and second in session two—a buzz that sounds like the cellphone interference people would hear through external speakers when a phone rings). You will notice that this type is not as markedly common in this session (noted only twice). At a few points in this session, a loud multitude of simultaneous indiscernible events is noted. This event is extended in length, and imposes a sense of competition between the sounds in question. Towards the end of this session, a similar type of events occurs, though some of the sounds can be identified: intense high-pitched loud scratches, and intermittent whistles/beeps that are of a rhythmic nature. By the end of this session, you will have noticed that it also features repeated sounds that change in volume and intensity, which determines whether or not they are notably heard.

HCMC Session 4[30]

During this session there were thirteen people in the lab space, two of whom were in the small cubicle, with a third person coming in at minute nine. The devices running in close proximity to the Detektor were: a laptop, a screen, Bluetooth headphones, and a cellphone. The Bluetooth mouse was intentionally turned off for the entirety of this session. You will hear that the sounds in this session are clearer and less crowded than previous sessions; it seems that a layer of sound—predominant in previous sessions—is missing in this recording, which is allowing for clearer results. However, this clarity you experience does not guarantee a straight forward and individual identification of sonic events, as you will notice. Further, similarly to the previous sessions you have listened to, session four has events that are dominant throughout the recording (buzzes, whistles, and hums) and others that occur occasionally and are disruptive in their nature. The first type of events is multiple buzzing sounds that are limited in length (one second long each). The second type of events is an intense static buzz, similar to the first type of events in session one (second type in session two and fifth in session three). This type notably reoccurs approximately four times. The third type of events is an intense high-pitched buzz, which is also common in previous sessions. You can count this event approximately three times, and will notice that it is sometimes interrupted by scratches. The fourth type of events is a change in volume and overall noises, where one can notice a drop in both pitches and frequencies. The fifth type of events is a louder scratchy staticky buzz. This event reoccurs around two times. As you will hear, this session features numerous interruptions in the overall rhythms and patterns of its sonic events.

HCMC Session 5[31]

During this last session there were six people initially, and nine people by the time the recording was finished. The devices running in close proximity to the Detektor were: a laptop, a screen, a Bluetooth mouse, Bluetooth headphones, and a cellphone; however, I stopped using the mouse around minute eight to inspect whether or not its movement was affecting sounds, so you might hear the difference there. This session is louder than session four. Similar to all previous sessions, session five has events comparable in nature to previously described events. The first type of events noted in this session is a three-part buzz, with the third part more extended than the first two. This event reoccurs once, as far as is clearly audible. The second type of events is similar to the first, but is a two-part buzz as opposed to having three parts. As you listen, expect for at least five occurrences of this event, and listen for how it is sometimes disruptive of others. The third type of events is a brief intense buzz that is 0.5 second long. This event is more common in this session than in others, and reoccurs around twelve times at different volumes and intensities. The fourth type of events is a rhythmic high and low of buzzing, with which scratchy and staticky buzzing interferes. This event reoccurs at a particularly crowded interval in the session. This session also features the whistle that you will have also heard during all other sessions. It also has some exceptional and interesting events that are different from other sounds (even across sessions), which are nonetheless difficult to individually discern or accurately describe.[32]

Causal Listening

In this section, I cautiously discuss potential causes that may have contributed to the variations of sounds identified in the reduced listening descriptions. Given the capacities of the Detektor, I assume that the EM waves transduced fall within the ranges of GSM networks as well as Bluetooth networks (within appropriate ranges). However, I do not completely exclude the EM waves from Wi-Fi network, as I do not have definitive evidence that they completely fall outside of the EM range captured. Further, given the multiple layers of sounds resulting in the recorded sessions, I argue that attempting to associate particular sonic events and rhythms with particular technological events would be unreliable. The technological events that I am aware of and noted in the setting descriptions are limited to my personal devices, because: a) requesting reports on colleagues' workings and technological actions during the times of recording would be a breach of privacy; and b) there are many factors that seep through the recordings that are completely outside of my knowledge, including but not limited to the GSM network and other EM waves emitting devices that are not considered within the scope of this research.

Nonetheless, there is one factor that I am capable of commenting on. As noted above, one of the devices that I was using at the time of the recording was an Apple Magic Mouse, connected to my laptop via Bluetooth. Listening to the recordings, I had suspected that the traffic from the mouse may be overwhelming other sounds. When recording session four, I turned off the mouse. As per the reduced listening discussion, session four has clearer sounds, though not

---

[32] I do not claim to have accurately described any of the sounds, given the complexities inherent thereto, as well as the intricacies of communicating sound through spoken and written languages.

completely discernable. The difference in volume and layers between sessions one, two, three, and five, and session four suggests that the Bluetooth connection is louder and more dominant than other connections. Although I have taken note of some Whatsapp messaging events during the recordings, I have not noticed any particular sonic events that are unique to the messaging process or consistent with timing of the recording, therefore I will not make assumptions in this regard.

Resulting Ambiguities

The Detektor—and Elektrosluch in some instances—allows the transduction of EM waves produced by technologies and network communications otherwise inaudible and inaccessible to people. Practicing reduced listening at the initial stage of analysis allows a deeper understanding of the phenomena at hand, if only by paying attention to the intricacies of sounds, and the complicated nature of communications processes so abstract. Causal listening adds a layer to understanding sounds resulting from these silent events; however, as demonstrated above, causal listening fails in addressing the multitude of information, even those collected in the process of describing what the listening devices transduce. Semantic listening proves to be a nearly impossible task, given the lack of symbolic language in network communications, as well as the absence of differential material that would be essential in understanding the meaning of transduced sounds, as communicated between networked machines. That said, the incongruities in detail between reduced listening and causal listening result in various ambiguities that are problematic when attempting to study how technologies communicate within an infrastructure. These ambiguities also result from the multitude of sonic events that seemingly correspond to different types of connections, as showcased in the case studies, without knowing what sounds correspond to what technological events: for example, we cannot be sure of what sounds are particular to GSM versus Wi-Fi—if any.

When listening to individual electronic devices, drawing connections between reduced and causal listening is a more or less achievable task, given the fact that the sounds we are listening to are mostly coming from one device. That said, the accuracy of analysis can be dismissed, due to the amount of activities and events that take place in order for an electronic process to happen. Although such events are interesting and raise their own questions, they are not of particular interest to this project, because the focus is technological networks and the rhythmic events of the communications inherent to them. As Miyazaki explains,

> rhythmanalysis of wirelessness would examine the highly technical processes and
> rhythms happening in those agencements of information by conducting a media
> technologically enhanced rhythmanalysis and by creating a systematic ordering of the
> noises, beeps, blips and pulses . . ., not only [by] listen[ing] to them, but also . . .
> explain[ing] their becomings (518).

A rhythmanalysis of network communications requires an understanding of the information responsible for the happening of a given connection.

Further, it is essential that a rhythmanalyst has access to the "systematic ordering" of the particular sounds produced, to attempt a proper explanation of "their becomings" (Miyazaki 518). The problem with such an exercise is the fact that the listening device available at the time of this study transduces a wide range of frequencies, which is too broad for its purposes. As per the AD8313 (the transducer chip installed in the Detektor) data sheet (*Analog Devices* 1998-2015), the wide bandwidth covered by the Detektor is 0.1 GHz to 2.5 GHz. Accordingly, it is rather difficult for one to determine what particular sound corresponds to what particular network, when the layers of sonification cover a multitude of networks simultaneously connected. It is also worth noting that the most advanced Wi-Fi technologies—currently on the market—function at a 5 GHz frequency range, which means that they remain outside of the transduction ranges of the Detektor. Given the prominence of Wi-Fi communications in the wirelessness of everyday life, leaving its EM waves outside of listening exercises possibly leaves out the most dominant traffic of network communications, and the most condensed of networks.

Peter Krapp argues that "[w]hen recurring noise patterns become signal sources as their regularity renders them legible, the systemic function of distortion doubles over as deterioration of message quality and as enrichment of the communication process" (xvi). Correspondingly, noise and noise patterns gain significance—signal—when they become regularly repetitive, whereby failure—distortion—simultaneously affects the message as well as enhances communication. Reduced and causal listening, unlike semantic listening, do not attempt to make (human) meaning out of given sounds, but focus on apprehending—and eventually comprehending—the communicative process between technologies and amongst networks. If a listener is to cognize sound as noise until it acquires a pattern, they risk missing the brief yet significant technological happenings. However, having a crowded sound clip that is rich in both event points and event patterns, without a clearly organized network organization, confuses the output between sound and noise, and thus contributes to fictional—and possibly forceful— construction and attribution of meaningful signals. In other words, a rich sound clip does not necessarily suggest enhanced network communications, particularly because the range of transduced frequencies and their layering may induce patterns that are naturally corruptive to our understanding of the process.

The multiplicity of sounds and their collisions could be collapsing sound and noise, which defeats the purpose of reduced listening and interrupts causal listening. Notably, the presence of a multitude of signals is essential to wireless communication. Comparing wireless communication to wired communication, Adrian Mackenzie explains that "[i]n wireless communication, nearly all signals are marked by the presence of other signals. The situation is overwhelmingly relational in comparison to the relatively narrowly constricted flows of networks" (Mackenzie 77). Thus, based on the indispensable yet non-constricted signal traffic, I argue that attributing individual causes to individual sonic events is irrelevant in open environments; rather, an exposure to the acousmatics of an infrastructure allows for a sensible access to the rhythms inherent to the multitude of algorithms, which are at play—power play for one—in making network communications possible.

Conclusion

To perform reduced listening is to listen with the intent of creating words to talk about sounds, describing what we can hear, with as much detail as possible, without taking into consideration the causes behind the sounds produced. On the other hand, causal listening requires the listener to identify particular causes behind the particular sound events. As demonstrated above, performing reduced listening to sound clips of transduced EM waves results in interesting and exciting findings that lead us to expect a lot of information. However, when attempting to exercise causal listening onto the same sound clips, particularly those involving higher frequencies and waves of network communications, it remains difficult to identify exactly what sources are causing what sounds. Semantic listening remains impossible at this stage, especially given the lack of knowledge that people have in making meaning of the transduced waves exchanged during streams of network communications.

The many layers of sounds and the various simultaneous rhythms are the result of the wide frequency range that the Detektor captures and transduces. Despite providing an idea of what network communications sound like in the ambient environment, the sonic congestion of the clips results in many ambiguities and raises various questions about the differences between the various types of networks. Exploring networks as a system of algorhythmic interactions allows people to approach network communications as a series of events, as they translate computational rules into vibrations that travel our ambient environment. These rules and algorithms are defined by a logic of power and control that not only regulates how the technologies we use and on which we rely communicate, but more importantly how *we* are to interact *with* and *through* them. As we hear them, our technologically realized actions produce and reiterate rhythms that redefine the natural space as inhibited by ubiquitous networks.

Chapter 4: Anti-Environments towards Multimodal Network Stories

Technological devices materially communicate via electromagnetic (EM) waves that translate mathematical commands from network algorithms to physical vibrations. The physicality and material nature of these communications processes is inaccessible to people, because the exchanged EM waves are at the level beyond the human hearing or sight senses. The intangible nature of communications endorses a culture of mystery and magic around algorithmic behaviours and technological networks. Sonification produces layered sounds and noises that compose a representation of EM ambient environments, conveying a physical and rhythmic materiality of the space within which devices of networks exist. I argue that the inseparability of the various rhythms and sounds is an essential if not intentional feature of network communications, one that guarantees a power and control dynamic, not only between machines and people but also between the machines themselves. In this chapter, I explore how listening can contribute to reducing the control that networks impose on users and the general public, by producing what Marshall McLuhan calls "anti-environments" that orient people's attention (McLuhan and Fiore 68) to networks as a system artifact.[33] In doing so, I examine how tactical network sonifications (TNSs) can act as a sonic medium that contributes to making network complexities more apprehendable—or even comprehendible—to audiences who are not technologically savvy.

---

[33] It is important to acknowledge that Marshall McLuhan's work can be problematic on various level especially through his colonial and universalizing language. As Sarah Sharma eloquently puts it in her editorial piece to the "Many McLuhans or None at All" special edition of the *Canadian Journal of Communication*: "[McLuhan's] terms espouse a worldview of the singular effects of media upon a universal subject—an attitude feminist media and critical race studies have long challenged" (484).

According to Tara Rodgers (2018), "[v]ibrations—including that specific class of audible vibrations experienced as sound—present alternative ways of apprehending reality that can point toward political sensibilities that emphasize complexity, interconnection, and interdependence rather than modes of distancing and control" (234). Listening to sonic artifacts can bridge between political complexities inherent to a system and the users that these convolutions implicate, by bringing forth the reality—and materiality—of the system's infrastructure. It is important to understand the complexity of algorhythmic sonic layers, as well as the interdependence of the factors that contribute to the sonic structure of a given space and time, because the layering of algorhythms is likely to affect and be affected by the soundscape within which a network operates. Going back to the dial-up example described in the introduction, listening to the soundscape of the environment can alert a person to, if not prevent an interruption: if someone is trying to connect to the internet and another person mentions that they are about to make a phone call, the person connecting would stop the other from proceeding with their call, so as to avoid interruption. To listen to an ambient environment, one would lay out the multiple realities that may be partaking in the structure of the setup—or environment—in question.

For example, the recordings I conduct in the HCMC, as described in chapter 3, have their own sets of complexities, mostly caused by the very nature of a semi-controlled environment: the facts that we cannot completely identify the exact number of individuals at every moment, and that we cannot track every machine or technological event, are examples of how distancing is not necessarily an end of the sonic approach, but more a condition of the nature of the question that this work addresses. More specifically, because the recordings are concerned with naturally non-tangible network issues (namely EM transmissions achieving network communications), the results do not actually distance us from the subjects. In fact, the audio files allow for a closer and more directed attention to the intricacies of the streams of communications happening within an environment, particularly because of the overlap between the many network—sonic—layers. Separating these audio layers is not sonically—or even mechanically—achievable. Deep listening provides an approach for investigating algorhythmic soundscapes as captured,[34] potentially "discover[ing] the significant features of the soundscape, those sounds which are important either because of their individuality, their numerousness or their domination" (Schafer 100). That is to say, deep listening is an exercise in identifying moments of sonic assessments, contributing to a process of soundscape analysis either via listening for events, or through an evaluation of a system's acousmatics. Listening to the algorhythmics of the networks engages the politics of algorithms with the politics of the everyday, especially when these sonic artifacts are socially and culturally situated as anti-environments to the constructed networks that our everyday lives have become.

Algorhythmics of "X-Reality"

Beth Coleman explains that "x-reality" is "[a] continuum of exchanges between virtual and real spaces. Pervasive media use defines a world that is no longer either virtual or real but representative of a diversity of network combinations" (188). Ubiquitous computing and pervasive media have allowed for a constant state of in between the "real" and the "virtual" to the point where x-reality has become the new norm for the everyday life. Coleman (2011) speaks

---

[34] Pauline Oliveros (2005) defines deep listening as a way of listening that "include[s] the whole space/time continuum of sound" (xxiii).

of x-reality in the context of pervasive media, which by definition is reliant on networks.[35] In my

research, I adopt her term and extend its scope to encompass the physical and material real

spaces within which EM waves roam as a result of networks; in doing so, x-reality is no longer

restricted to people using the machines, but also includes machines communicating with and

through each other. The difference between x-reality and reality is that the former includes the

nontangible yet material presence of network communications, whereas the latter is only

concerned with the tangible aspects of networking technologies, particularly those of which

people are aware. In listening to networks and their communications, the physical and material

realities of networks merge with their virtual world, also known as the mysterious cloud; x-

reality becomes the outcome environment of "spatio-temporal rhythms of nature as transformed

by a social practice" (Lefebvre 117), such as networks. While networks connect devices and

people in an invisible state, transducing the EM waves that make these connections possible

produces a layer that people can actually hear and virtually see—via sound visualization

software for example.

Given that algorhythmics allows scholars to listen to the algorithms responsible for

technological events, the algorhythmics of x-reality make its very scope possible (as developed

here). Correspondingly, using the Detektor as an algorhythmic device captures the physicality

and materiality of network communications, allowing for a translation and understanding of the

virtual into a more sensible physical reality. In listening to the rhythms that a system entails,

---

[35] **Pervasive media:** "A global culture that engages a spectrum of *networked technologies*. Platforms include virtual worlds, voice-over-Internet protocol (VoIP), mobile rich-media and texting, and microblogging formats. Implicit to pervasive media engagement is a convergence of multiple transmedia forms. The term borrows from the language of computer science where pervasive computing, also known as ubiquitous computing, describes a world in which objects, places, and gestures are included on a *computational network*" (Coleman 188; my emphasis).

algorhythmics can also redefine the time and space of x-reality. Given that "[r]hythm is the order of movement, [and] timing of matter, bodies, and signals" (Miyazaki 129), understanding the rhythms of x-reality raises the following questions: The timing of what matter would be investigated? What/who are the bodies involved? What/who is emitting the signals? Upon first and superficial investigation, one can say that the matter to investigate is EM waves, and the bodies involved are networked technological devices that in turn emit the signals in questions. The human factor may seem to be unimportant or irrelevant to the subjects of interest, especially since people's role in network communications has become either deep in their process, or taken for granted as outside of a network's functioning (until a problem occurs and blame is be assigned, as discussed in chapter 2). The algorhythmics of x-reality reintroduce the human factor to the network environment if merely by acknowledging the role of people at the early stages of network communications and network design. The approach also establishes a means for people to participate in the conversation *on* network environment, by listening to that *of* the networks. More importantly, the algorhythmics of x-reality engage a sensible reality that extends beyond the role of people, to re-examine the time-space continuum *across* as well as *within* which pervasive media and networks oscillate and balance.

Coleman (2011) argues that in the context of pervasive media, "participants [people] must see themselves as agents in a way that informs the situated nature of a system, as opposed to having our objects inform us" (158). Passively accepting a course of action within a pervasive media system prevents people from exercising any form of agency; thus, it remains essential that people actively take on a responsibility to participate in a system, in an effective and intentional manner. Further, Wendy Hui Kyong Chun (2011) explains that "[o]n networks, the agents seem to be technology rather than the users or programmers who authorize actions through their commands and clicks. Programmers and users are not creators of languages, nor the actual executors, but rather living sources who take credit for the action" (102). Even though somewhere deep in the layering of network communications, the human factor plays (or more accurately played) an essential role in the design and making of the process (including hardware, software, the physical, and the virtual), people remain at the peripheries of x-reality. That said, it is not always the case that people are passive participants; technologies are often design to function without even allowing much critical practices on behalf of users, who in turn may be using a given technology out of sheer necessity. Additionally, the effective functionalities coupled with the complexities inherent to networks, make it rather challenging for people to even want to be active participants in a system, beyond their mere roles as users. To arrive at a networking system that allows for people to be participants in an informed way, it is essential to invite them to explore—or even simply recognize—the complexities of objects at hand, even if *by* the objects themselves when necessary. I argue that the algorhythmics of x-reality—through TNSs—are first intended to embody the complicated physical reality of network communications within their infrastructure, to reintroduce the human factor into the broader equation.

Shintaro Miyazaki (2018) explains that "algorhythmic sensitivity allows people to experience and understand structures of a wide variety of key media operations, their fundamental principles, and their timings. This implies a sensorial, nonlinguistic approach to the inner workings of computational gadgetry" (246). Listening to the rhythms of technological devices allows people to hear the manifestation of technological operations, within a given timeframe, thus accessing at least one aspect of their functional standards: algorithmic temporality. By listening to the algorhythmic operations of pervasive media through their corresponding networks, people gain access to a mode of information that was previously restricted to the machine. Interestingly, algorhythmics of x-reality not only sonify communications between technological devices and the corresponding network algorithms, but also make sensible timed automatic changes and updates that the machines produce and generate, outside of human intervention. Coleman (2018) explains that "an IoT interaction is . . . based on the conditions of pervasive media in which devices talk to each other, automatically updating agendas, programs, scripts, and so on—creating a network of machine-to-machine (M2M) communication" (223). To put differently, the process of autonomously developing itself within its network is an essential characteristic of a network device, and an important function towards network communications.

Given that "IoT interaction[s] [are] fairly *invisible*" (Coleman 223), it is difficult for people to evaluate IoT communications processes within their physical reality and away from virtual representations. Nonetheless, "[t]he problematic relationship of algorithms [and consequently networks] to reality is mediated by signals of mostly EM waves going through wires, air or another medium" (Miyazaki 129). Making accessible to people what was once only machine-to-machine communication, algorhythmics opens new streams of information that are not textual or necessarily visual, presenting as an expression of the dialogue within our network environment. Introducing participants to the intricacies of networks—especially ones that maintain pervasive media—paves the way for them to make informed decisions beyond mere assumptions and calculations, when they are so inclined. That said, sonifying network communications blurs the lines between the real and the virtual, even if bridging between the two, and makes sensible network processes—and their rhythmic inscriptions on the natural space (Lefebvre 1991).

Miyazaki (2018) explains that "[t]he growing ecosystem of intelligent machines and small invisible devices, which are connected to our smart phones, tablets, and laptops, generate[s] a never-ending stream of algorhythmic effects that may influence processes on a planetary level" (248). In other words, the fast pace at which technologies (particularly pervasive media) develop has an invasive potential not only at the level of human-life, but also extending to other species inhabiting the Earth. The EM waves responsible for the functioning of these technologies can have various effects on animals, plants, and living organisms that respond to these waves in a more recognizable and tangible way than humans do—in addition to other planetary risks attributed to artificial intelligence for example.[36] Networks and the IoT are generous in providing a constant—though not necessarily consistent—stream of rhythms that (re)generate a time-space continuum at the intersection of the physicality and virtuality of the technological realm. Applying algorhythmics to x-reality entails an intricate examination of the interrelationships between physicality, virtuality, time, and space. More specifically, the connection between time and space as determined by the algorithmic events underlies the connection between physicality and virtuality.

The high number of IoT devices, for instance, holds relevant in that it contributes to the crowded recordings, as discussed in the previous chapter—"Listening for Network Events." The crowded sound recordings suggest rich time-space continuums of networks, given the multiple layers of rhythms produced. Tim Edensor (2018), in reference to Henri Lefebvre's work, explains that "the multiple rhythms produce [an] ongoing spatial fluidity" (158). The multiplicity

---

[36] Training AI models produces a very high carbon footprint. Although this particular issue remains outside the scope of this study, the risks are worth mentioning. For more on this topic, see "Energy and Policy Considerations for Deep Learning in NLP" by Strubell et al. (2019).

of networked devices and their communications traffic propagate their events and beings across a spatio-temporal realm of the ambient environment within which they are bound. Tactical sonifications of the algorhythmics of x-reality can redefine and reshape the relationships between networks and time and space of the everyday.

Lefebvre (2013) explains that "[e]verywhere where there is interaction between a place, a time and an expenditure of energy, there is **rhythm**. Therefore: a) repetition . . . b) interferences of linear processes and cyclical processes; c) birth, growth, peak, then decline and end" (25). The rhythmicality of network communications and the movements between the physical and virtual of x-reality, together inform an everyday life, situationally redefined to oscillate on a y axis of machine-to-machine (M2M) interactions and communications and an x axis of human to machine interactions, both axes extending between the physical and virtual. The y axis represents the M2M communications. Point zero of the y axis represents the physical aspect of x-reality; the higher the axis goes, the more virtual the so-called reality becomes. Further, the higher one goes on the y (M2M) axis, the less agency the human factor has, the more autonomous the technological interactions become. The x axis represents the human-to-machine (H2M) interactions. It shares the same points for physical (at point zero) and virtual (the more one progresses on the axis). In contrast to the y axis, progressing on the x axis does not add or remove agency to and from neither humans nor machines. The progression, however, makes the (H2M) interaction more virtual. Take for example a phone call between person A, who has an Apple device, and person B who has an Android device.

At the very beginning of the call, A calls B. At this point, the events on the x axis are: A dials B's number, communicating to their machine the command to connect with B. Next, on the y axis, the events are as follows: A's device connects to the network (event one), and requests (event two) that the network connects A to B (event three). All three events require a physical transmission of EM waves; however, the process of decoding happens at the machine level, in a virtual realm to both A and B. The oscillation between physical and virtual has begun on both M2M and H2M levels, whereby A sent one command leaving their machine to proceed into a virtual process (with respect to the person in question). Once A and B are connected and chatting, M2M is continuously sending, decoding, receiving, and decoding, the EM waves that maintain the connection. On the y axis, the interaction is continuously oscillating between physical and virtual, getting minimally in touch with A and/or B as they send their messages (i.e. speak). However, the H2M interaction (x axis) becomes more virtual, as A and B communicate with each other, without a conscious recognition of their interaction with the machine itself. The rhythmicality of network communications consolidates rhythms of x-reality that adhere to a sense of stability, though constantly evolving to adapt to their environments.

The pervasiveness of networks, including their physical aspect (the EM waves), requires that the rhythms of x-reality adapt to and integrate in the various "real" and "physical" environments of our everyday life. Coleman (2011) explains:

> With X-reality, I mark a turn toward an engagement of networked media integrated into daily life, perceived as part of a continuum of actual events. This is a movement away from computer-generated spaces, places, and worlds that are notably outside of what we might call real life and a transition into a mobile, real-time, and pervasively networked landscape (20).

The time-space continuum of pervasive media has become embedded in the "real" everyday network, after evolving beyond a state restricted to computer generated-spaces (VR worlds for example). The totality of a network is more or less defined by the range of EM reach that the waves can establish. Nonetheless, the interconnectedness of the various communications methods across devices makes for a more complicated spatial recognition of a network. Further, with pervasive media, it becomes difficult to separate networks into individual and distinct channels of communications, at least in terms of EM waves. Re-examining the time-space continuum of x-reality as per algorhythmics goes as far as considering the IoT networks as a system. In that sense, the "pervasively networked landscape" (Coleman 20) is analogous to a pervasively network soundscape. This soundscape is a reproduced sensible *layer* of the everyday, which in itself adds to the complexity of x-reality. That said, the rhythms of x-reality may or may not coincide and agree with those of the audible everyday life, while still redefining the natural space that we live in.

Lefebvre (2013) claims that "the everyday reveals itself to be a polyrhythmia from the first listening" (25). The rhythms of everyday life together with those of network communications, though they do not constantly and consistently harmonize, mostly exist in a polyrhythmic state—that is not to say that x-reality does not ever have arrhythmic instances. The polyrhythmic state of x-reality also feeds on "a sector of X-reality that purposefully exploits the experience of intersecting levels of information, engagement, and agency" (Coleman 147). Information, engagement, and agency also contribute to a fluctuating time-space continuum in that they operate within different levels of rhythmicality on the axes of M2M and H2M: "Time and space without energy remain inert in the incomplete concept" (Lefebvre 70): the time-space continuum is activated by the energy that would allow for such fluidity. Considering the EM waves to be the energy that activates x-reality, they become the energy that "animates, reconnects, and renders time and space conflictual" (Lefebvre 70). The rhythms of EM waves are a sensible means of studying x-reality's time-space continuum, investigating how the continuum itself enhances or disrupts the flow between the physical and virtual. Not to forget, the TNSs involve overlapping networks, which could result in conflict and/or arrhythmia that simply envelops the system in question.

Power and Control as We (don't) *Hear* Them

Edensor (2018) explains that "[i]nteraction with space is . . . never solely subject to symbolic signification but also to an embodied knowledge partly constituted by a sensual understanding deepened by time and embedded in memory" (159). Experiencing space in meaningful and affective ways benefits from embodied lived experiences and sensible interactions with time, within physical and intellectual memory components. People are exposed to an embodied sense of rhythm as imposed by networks, especially when they share a space with pervasive media—physical or virtual—throughout extended and consistent periods of time during which technologies communicate. Michelle Duffy, Gordon Waitt, Andrew Gorman-Murray, and Chris Gibson (2011) claim that "[t]he porosity of our bodies means we also *feel* sound waves that we then comprehend and (re)constitute as a pulse, as a rhythm" (18). Our bodies allow for us to absorb sound and vibrations, and (unknowingly) re-enact them through our own physical rhythms and pulsations. A tangible example is *feeling* the bass of a song, through the vibrations emitted by the speakers, particularly when the bass is set to high. Similarly, though less noticeably, people are susceptible to *feeling* the EM waves as generated by network communications, and to processing them as embodied rhythms. Algorhythmics allows people to investigate the sound waves that our *bodies* would eventually comprehend. These EM waves travel in what is called the "air interface" (Mackenzie 66). Adrian Mackenzie (2010) explains that the *air interface* is a term for that part of a wireless or mobile telephone network that lies

between the antennae of a device and a base station. It is an elusive interface, one that

shows no face apart from the tips of antennae and the more or less conspicuous towers

and masts of telecommunications and telephone service providers. The air interface,

however, is synthesized by technical processes expressed in algorithms. The algorithms

generate waveforms that support conjunctive pathways (66-7).

The air interface is naturally invisible; people can comprehend it through the mere existence of

what we might refer to as the physical nodes of networks—antennae and towers for instance—

imagining air particles to be the edges that connect these nodes to each other and to our devices.

The communications that happen between our devices and those that arrange

communications on more global levels (like routers < antennas < towers), redesign the invisible

architecture of air, as per the architectures of the algorithms responsible for telecommunications.

Further, the air that envelops us becomes but a path that information travel to establish or

maintain a connection. That said, people are bound to become objects with which EM waves

engage as they travel the air interface. Consequently, while unknowingly subjected to said

waves, we subconsciously *feel* and embody not only waves as rhythms, but more importantly the

algorithms that architect air as a techno-communicative space. Not to forget, the techno-

communicative space itself is not foreign to x-reality; in fact, it is a main component of its

physical infrastructure. In listening to the soundscape produced by the x-reality of an

environment, people may identify patterns that intercept with the everyday, if only by invading

the air we breathe without our consent or even awareness. Tactical network sonifications allow

people to physically experience how network communications might regulate their rhythms,

through the porosity of their bodies. Further, listening to how technologies communicate within the polyrhythmia and/or arrhythmia of x-reality, expresses a form of power and control fed to our bodies in most discrete ways, via architectures of algorithmic control (Lessig 2000).

Network protocols and algorithms regulate and control networks by setting rules and standards that allow for communications to happen across various platforms (Galloway 2004). The diversity of technologies and communications environments requires regulated means of data transfer, which is not restricted to software but also translates to hardware (Galloway 7). Additionally, the various communications channels and streams require regulation, *within*—as opposed to *across*—technologies. More specifically, devices that operate via Wi-Fi do so as per protocols and rules unique to Wi-Fi communications, as opposed to GSM communications for example. These protocols, like other communications and technological protocols are realized via algorithms that control the traffic of EM waves, and subsequently the air interface. Not surprisingly, algorithmic control and governance is enhanced—if not augmented and automated—via prominent networks and corresponding advancements of infrastructures.

Claudio Coletta and Rob Kitchin (2017) explore how "an assemblage of related governmental technologies . . . rather than individual algorhythms" allows for an understanding of "urban governmentality enacted by algorithms" (5). Investigating an algorithmic assemblage not only takes into consideration individual algorithmic control, but more importantly highlights the technological power imposed within our environment when various algorhythms work polyrhythmically or generate arrhythmia within a system. I argue that polyrhythmia and/or arrhythmia, as investigated through the tactical sonifications of the algorhythmics of x-reality— and as sonically experienced in chapter 3—manifest power and control as *performed* by technologies competing to dominate the air interface in their communicative processes. For example, Bluetooth communications may need to override or "speak louder" than some Wi-Fi technologies as they communicate within approximate frequency ranges. Arguably, as discussed in chapter 2, media platforms attempt to participate in technological power dynamics by interrupting the dialogue between the various participants within a network story. In doing so, they risk leaving the general public—audiences who are not necessarily technologically savvy— a rather powerless and controlled subject at the peripheries of sociotechnical stories.

Kitchin (2017) urges that "it is most productive to conceive of algorithms as being contingent, ontogenetic, performative in nature and embedded in wider socio-technical assemblages" (16). Investigating algorithms within their environments, provides a more or less holistic approach to understanding their existence in our everyday in a sensible way. Although listening to algorithms does not necessarily allow for an understanding of the technical complexities of individual networks per se, it still allows access to "algorithmic authority" (Kitchin 19). In other words, listening to networks is a way around one of Kitchin's "significant challenges to researching algorithms": the fact that they are "heterogeneous and embedded" (20).[37] As he explains, being able to investigate the technical intricacies of an algorithm, including its mathematical set-up, its logical rules, and its functions, "will not necessarily provide full transparency as to its full reasoning, workings, or the choices made in its construction" (21). Mike Ananny and Kate Crawford (2018) explain that transparency is not the simple answer to algorithmic accountability. In fact, the two authors call for "hold[ing] systems accountable by looking *across* them—seeing them as sociotechnical systems that do not *contain*

---

[37] In "Thinking Critically about and Researching Algorithms" (2017), Kitchin identifies the following "significant challenges to researching algorithms": "Access/black boxed;" "Heterogeneous and embedded;" and "Ontogenetic, performative and contingent."

complexity but *enact* complexity by connecting to and intertwining with assemblages of humans and non-humans" (974).

Investigating systems would benefit from people studying them collectively, to understand their effects on the environments around them. Part of such investigation requires an understanding of how these systems together perform their intricacies within the contexts within which they function and thrive. Tactical network sonifications fail at untangling network communications into separate network streams. They also fail at providing access to individual algorithmic complexities outside of their networks. The limited access to the semantics of tactical sonifications is mostly due to the lack of linguistic or even symbolic language in network communications. However, this limitation was productive in its own right: it directed my attention to the importance of evaluating the infrastructure as a collection of networks. The TNSs present themselves as a methodology, not only for inspecting algorithms in an accessible way, but more importantly for investigating them within their network performances. In their polyrhythmia—as well as arrhythmia—and through EM emissions, networks enact corresponding algorithms as per a dynamism reliant on the exchanges between humans and non-humans of a given assemblage—in the various possible directionalities, and through the multitude of communications paths: human-human, machine-machine, machine-human, human-machine.

Chapter 3 provides extensive sonic examples of the complexities of network communications at play, available to the public on an open platform. Although the consistency in the semi-controlled environment approach (as described in the previous chapter) may suggest that the results would be more or less coherent, the recordings were considerably different from each other. In addition to the factors that were controlled (space, time, duration) there are various factors that were not and could not be monitored, among which is the presence of people in the space and the work that they are doing within the corresponding space-time continuum. The sounds, patterns, and rhythms that were common and shared between the five sessions suggest that some technological activities are consistent and repetitive. However, the changes in in the various sonic characteristics, including but not limited to volume, frequency, pitch, and pace, hint at a dynamism beyond the consistency of connection maintenance. Further, the drastic differences between sessions highlight the important role of people in the assemblage recorded. That said, human-machine as well as machine-machine communications come into play to design the soundscape within which people and machines operate. Although people and machines may interact as agents, their respective agencies are not necessarily equitable, in that algorithmic power overrides and controls the contributions that people make to a certain soundscape.

As per the recordings, people's interactions with technologies can affect features such as volume and pitch, in a rather unintentional way. That is to say, even though some of the actions that people perform may contribute to the sounds produced, I argue that people are not responsible actors in the assemblages unless they behave with the particular intention of altering the sonic outcome. People's active participation is tied to their ability to identify at least what technologies sonically dominate, and whether or not they do so beyond the soundscape. For example, if operating a Bluetooth mouse sonically dominates over other technological sonic events, does it also affect the system in question, or is it only producing louder buzzing that ostensibly overrides other algorhythmic processes? More so, can people's interactions with the machines prompt a certain algorhythmic behaviour that enacts notions of power and control? Tactical sonifications of the algorhythmics x-reality may not necessarily empower people to override the governance of network protocols, but they make sensible the depth and layers of the logic of network communications. In discussing transparency in the context of algorithmic accountability, Ananny and Crawford (2018) argue that "transparency is . . . a system of observing and knowing that promises a form of control" (975). That is, when we call for transparency, we are not necessarily encouraging an ultimate access to the technical details of an algorithm; what algorithmic transparency calls for is access to a form of information that allows people to participate in the happenings of algorithms, even if only through an understanding of the general operations of the systems in question. Tactical network sonifications is a potential method to approach Ananny and Crawford's algorithmic transparency; listening to the rhythms of algorhythms, especially in an IoT infrastructure, can allow people a sense of epistemological control, at least in monitoring algorithmic behaviour and listening to networks.

Tactical Network Sonifications as Anti-environments

Batya Friedman and Peter Kahn (1997) argue that in addressing issues of responsible computing, it is important to direct focus "not only on consequences of acts, but agency – on what and why some things can be held morally responsible for action" (221). Working towards a responsible computing system begins by acknowledging who is capable of intentionality (Friedman and Kahn 1997), to allocate agency where appropriate. Although humans design and implement algorithms, algorithmic processes have the capacity to learn, develop, and even behave without a direct contact from people or even their knowledge. Such capacities do not necessarily suggest that algorithms have intentionality; they do, however, raise various questions about agency. While I agree with Friedman and Kahn that computer systems cannot be "moral agents" (225), I argue that algorithmic agency—even if not moral—still needs to be mitigated, towards responsible algorithmic practices. Friedman and Kahn claim that "to support humans' responsible use of computational systems . . ., system design should seek to protect moral agency of humans and to discourage in humans a perception of moral agency in the computational system" (226). Accordingly, TNSs could be implemented to endorse both requirements, without restricting algorithmic agency to one of moral nature. In listening to the acousmatics of a system people can start actively and intentionally investigating algorithmic behaviour and infrastructure. Tactical network sonifications allow us to examine the extent to which an algorithmic system or a network is independent of human actors, and to estimate how much people contribute to the system's operations, in comparison to the automated functions of network algorithms.

McLuhan states that "[e]nvironments are not passive wrappings, but are, rather active processes which are invisible . . . . Anti-environments, or countersituations made by artists, provide means of direct attention and enable us to see and understand more clearly" (McLuhan and Fiore 68). Consider the x-reality of a given system to be an environment as per McLuhan's claim, and the TNSs of the algorhythmics of x-reality to be the anti-environment. Assuming agreement upon the notion that x-reality is an active invisible, inaudible, and non-tangible process, TNSs of x-reality begin a process of adjusting participation, by allowing people a sensible directing of attention towards a clearer understanding of the infrastructure. Further, consider Galloway's (2012) take on "information economy":

> We must speak of the information economy. We must simply describe today's mode of production in its many divergent details: the diffusion of power into distributed networks, the increase in local autonomous decision making, the ongoing destruction of the social order at the hands of industry, the segmentation and rationalization of minute gestures within daily life, the innovations around unpaid micro labor, the monetization of affect and the 'social graph,' the entrainment of universalizing behaviors within protocological

organization – *these* are the things that are unrepresentable . . . . The point of
unrepresentability is the point of power. And the point of power today is not in the image.
The point of power today resides in networks, computers, algorithms, information, and
data (92).

Power, as linked to information, has become inherent to the structure of networks defined by
algorithms and protocols. More importantly, power is disseminated and/or retained because of its
"unrepresentability," inherent, yet again, to "protocological organization" (Galloway 92).
Tactical network sonifications as the anti-environment provide a form of representation that
could disturb power structures, and adjust universalized (algorithmic) behaviours to fit within a
more just information economy of the everyday x-reality.

Given that "[p]roducers of the commodity *information* know empirically how to utilize
rhythms" (Lefebvre 57), allowing people access to these rhythms is allowing them control in that
regard. A redistribution of power relies on a redistribution of control, whereby granting people
access to network algorithmic infrastructures would not only give them an opportunity to be
informed, but more importantly add them to the equation of power equilibrium of networks, if
ever there was one. Moreover, Galloway (2004) predicts that "in the future one is likely to see
bilateral organizational conflict, that is, networks fighting networks" (244). When listening to the
crowded anti-environment soundscape that the Detektor recordings present, as explored in
chapter 3, one can safely say that the future is here: networks are fighting each other to dominate
an architecture of communications, designed to be inaccessible to people, especially those who
are not technological experts. The layers of sounds, the changing volumes, the increasing and
decreasing intensities of buzzing, are all examples of how communications algorithms are
competing to achieve and maintain their connections. Investigating TNSs as anti-environments,

people can monitor this competition and listen to a system that no longer manifests "hierarchical powers and distributed horizontal networks" (Galloway 244). A responsible network would embrace sonification to invite maintenance and development, and would accept algorhythmic anti-environments as an informative technique for its designers and makers, as well as its general public.

Shannon Mattern (2018) explores how sonification can be used as a forensic method and a technique for investigating mechanic and programmatic infrastructures. She explains that "[s]ound serves as a useful diagnostic tool" and that it is very capable in informing people about "the evolving processes of computation by listening to the internal mechanisms of these machines" (223). Giving people a diagnostic technique that is comprehensive and more or less exhaustive allows for a set of interventions that have social and cultural interests at the heart of their concerns, as opposed to the merely technical advances that the technological industry may be inclined to focus on. Rodgers (2018) explains that "we learn who we are as socially differentiated bodies and subjects in part through our engagements with sound and music. Because sound is felt by the body in complex ways, it holds particular power to move us physically, emotionally, intellectually, and politically" (236). In that sense, TNSs as anti-environments not only make sensible the power politics inherent to algorithms, but they also make it possible to understand them within the everyday political and social realm.

In a practical example, TNS can serve as a diagnostic technique to investigate infrastructural distribution across communities. In doing so, it allows people to question potential discriminatory network availability and operation, especially when comparing settings that are identified as controversial areas (in terms of social and cultural discrimination for example). Additionally, listening to an algorhythmic recording and documenting a personal response, be it physical or emotional, can manifest an intellectual or political response to the technological networks investigated. The political weight that TNSs carry is yet another way by which listening to the algorhythmic of systems can promote anti-environments as diagnostic techniques: "Media, by altering the environment, evoke in us unique ratios of sense perceptions. The extension of any one sense alters the way we think and act—the way we perceive the world" (McLuhan and Fiore 41). The way we think of and act upon networks of everyday is redefined by how we respond to their algorhythmics and sonifications. Consequently, tactical networks would consult their algorhythmics and/or sonifications to ensure a polyrhythmia—if not eurhythmia—that aligns with the social and cultural placings of the systems in question.

Conclusion

X-reality is when the virtuality and physicality of pervasive media meet and form a unified realm (Coleman 2011). The TNSs of the algorhythmics of x-reality present a sensible rendering of network communications, especially their material aspect. In doing so, they maintain people's informed access to x-reality. Given the complexity and interconnectedness of networks, TNSs of the algorhythmics of x-reality are more effective if addressing networks as an infrastructure, as opposed to focusing on the intricacies of a single network at a time. Investigating the acousmatics of such a system makes it possible for people to question the power politics between networks, as well as how they translate into the politics of people's everyday. Further, TNSs allow a form of transparency that could permit people epistemological control in the happenings of algorithms, even if beyond the technical specifics of a given algorithm. Because algorithms are inherently controlling, it is imperative that people have some form of participation that promises them an understanding of how control is affected within an infrastructure. Responsible computing demands that people have sensible access and that moral agency is not attributed to algorithms (Friedman and Kahn 1997). Tactical network sonifications act as the anti-environment of networks, and become a diagnostic technique for network communications processes. In other words, they, through the politics of sound, promote a change in how we respond to the networks within our environments.

Eventually, this practice may encourage a shift in people's behaviours, which in turn will not only achieve responsible use of computational networks, but also promote responsible and tactical network designs. For example, people may be inclined to investigate the network soundscape within their living spaces, to design them in ways that separate between productivity and leisure spaces. More specifically, in one's home, one might want to locate the Wi-Fi router in the office or work space, so that the dinner or communal spaces are limited in network infrastructure—preferably to the network communications that seep from outside of the house/apartment (such as GSM connections as opposed to Wi-Fi). To encourage such practices, tactical network designers might resort to TNS in testing the infrastructure within which their network would fit. In doing so, they can work towards a responsible network that accounts for the possibility that people are aware of the contribution of an added network to their soundscape. A radical change, in particular, would be a feature—a button or switch—in connectivity routers that transduces the EM soundscape as part of the installation and testing processes.

Conclusion

Networks are indispensable in today's everyday life, providing the infrastructure for many of the components that allow and support current communication streams, including but not limited to the Internet of Things, artificial intelligence, and algorithms. While people find it essential to engage in critical conversations around such a prominent factor of our daily lives, network stories are still defined by the magical aspects of efficient technologies—as per Arthur Clarke's Third Law (Clarke 1973). The stories that circulate around networks and their components often contribute to the culture of fear and mystification of technological advancements, to the point where misinformation and disinformation become more the norm than the exception. People are often lured to either fear and avoid networks, or simply accept them for what they are assumed to be: magical boxes that allow us to socially and professionally operate, as long as we do not question or interrupt their overwhelming processes. The fact that technology users do not need to be technological experts makes it possible for systems to be designed and implemented in abstract ways. Arguably, the magical performances of technologies are encouraged, to achieve the level of convenience which users might be seeking.

Networks, however, are material in nature, and are therefore physically established and maintained within our environments. Network communications rely on the transmission and reception of electromagnetic waves, to establish, maintain, and achieve their connections. These waves have a material presence, and travel in the form of vibrations between the nodes of a network—the devices that establish connections. Speaking of networks as though they lack materiality, and further removing the human role from their design and establishment only complicates how users may perceive such infrastructures, especially for those who are not experts. The fact that networks are complex and extremely technical, does not excuse telling stories about them in ways that abstract them from the public instead of engaging people with their intricacies.

Networks are political by nature, especially given their algorithmically determined behaviour. That said, the power and control inherent to network performances are aspects that should not be made discrete to the general public. In fact, obscuring such features of networks contributes to their power structure, and reinforces an infrastructure of control within which people are assumed to be oblivious consumers who live in fear of and submission to the dominance of the networks. This portrayal of passive technology consumers has been prominent in *The Social Dilemma* Netflix documentary (2020), as observed in the introduction of this dissertation. In telling stories about networks, media outlets assume a responsibility in confusing facts and aspects of the sociotechnical environment of the everyday life, as explored in chapter 2. Such confusions result in complicated relationships and tensions between the general public—especially technology users—and the designers and makers of networks.

Further, the public is rarely allowed informed reflections upon a given system, even less so when the role of people in the design and making of these systems is conflated with the autonomy and automation of the systems themselves. In other words, the relationship between the general public and networks is established on uninformed grounds that assume fictional abilities on behalf of the technologies involved, without anticipating the human factor that is essential to the very essence of networks; critical reception of network stories becomes difficult if not impossible. Approaching network structures as such, reinforces an unbalanced power and control dynamic that the technological sphere already assumes and on which it strives. To address the political nature of networks and begin to untangle the complexities of their established control over the sociotechnical aspects of everyday life, people would benefit from having the means to investigate networks beyond their abstract and ambiguous stories. That said, when people gain access to network events through their material executions, they can investigate the complexities of networks. In doing so, they experience them in sensory ways that enhance embodied awareness of governing constituents of the everyday life.

Marshall McLuhan says that "[t]he medium, or process, of our time—electric technology—is reshaping and restructuring patterns of social interdependence and every aspect of our personal life" (McLuhan and Fiore 8). The patterns of our social life have become deeply rooted in sociotechnical rhythms, given the prominent role of network communications. According to Henri Lefebvre, "there is a close relationship between rhythms and the wave movements studied in mathematics and physics" (87). EM waves responsible for the material connections between the many nodes of a network, are thus rhythmic in their nature. To attempt at understanding how technological networks may affect patterns of our social life, in this dissertation I have proposed listening to their rhythms and events. The idea of listening to networks builds on two important arguments: a) current media representations abstract the nature of networks and their physical presence in our environments; and b) addressing the material aspect of networks provides a sensible approach to network behaviour, and allows for an embodied experience of systems that are otherwise obscured and intangible—as explored in chapters 3 and 4. It is important to note that this dissertation positions the general public—people who are not technology experts—as its main audience: listening as a technique permits a flexible and non-elitist methodology for knowledge dissemination within Science and technology Studies.

Although the scholarship that determines the theoretical framework of this project's arguments is mostly situated within the academic sphere, the practices and techniques are inspired by and build upon various projects designed for artistic and experimental purposes, most essential to this project is Shintaro Miyazaki and Martin Howse's Detektor. In transducing EM waves, the Detektor permits a sonification of the ambient EM waves between 100 MHz and approximately 5GHz (Miyazaki 2013) that networks produce in the silence of their operations. These waves carry rhythms in their vibrations, and are thus representative of the "algorhythmics" of networks (Miyazaki 2013). In listening to these rhythms, people have access to sonic events that in turn represent certain events within network communications processes. Although I do not claim to decipher—or expose—what these events could possibly mean, having access to the sonic occurrences is a step towards a tangible understanding of the complexities of network operations. Sonifying technological procedures that are hidden from the public and explicitly rendered incomprehensible by the media is one method towards critically engaging with the network infrastructure of power and control. The sound recordings of a network infrastructure at the University of Victoria, discussed in chapter 3, are an example of how the multiple components of a single infrastructure are in constant competition and sonic overlap.

When listening to a network, people experience the EM ambient environment in an active and intentional way: TNSs of the algorhythmics of a sociotechnical environment (an x-reality) are the anti-environment (McLuhan and Fiore 1996) that could participate in breaking barriers between network stories and the general public. More importantly, they allow people an active role in listening to network infrastructures; in doing so, they reallocate structures of power and control beyond the networked objects and techno-elitist communities. Jonathan Sterne emphasizes that the term "audile" "refers to the physiological process-based sense of hearing . . .

and . . . references conditions under which hearing is the privileged sense of knowing or experiencing" (Sterne 96).[38] Listening is an active and conscious effort to use our hearing senses, to experience our surroundings in an intentional way, allowing for an acquisition of knowledge predominantly reliant on our physiological capacities. In physiologically engaging with networks, people are invited to reconsider their relationship with these infrastructures, rejecting the notion of inaccessible information, and learning to receive and interact with knowledge through embodied experiences. TNS emphasizes the role of listening in approaching technologies so often intentionally kept mysterious to the general public. Sterne explains that

> Audile technique was not simply a representation of acoustic space; it aimed actively to transform acoustic space. The space occupied by sounds becomes something to be formed, molded, oriented, and made useful for the purposes of listening techniques. It can be segmented, made cellular, cut into little pieces, and reassembled (93).

Recording the acousmatics of a certain network is an audile technique that prepares the acoustic space of the infrastructure for TNS—as a listening technique. The space in question—x-reality for example—becomes a subject ready to be studied through segmentation, reassembly, and other practices: listening for rhythms and events proposes that audiences work with the product as best suitable, to form an opinion as to the system in question.

Tactical network sonifications give people an opportunity to listen to a network on repeat, in pieces, or in loops, encouraging an apprehension—and even comprehension—of an infrastructure through its sonic events. Further, TNSs allow a flexible practice, whereby people

---

[38] Jonathan Sterne defines audile as such: "An audile is a person in whom auditory knowing is privileged over knowing through sight. As an adverb or adjective, it means 'of, pertaining to, or received through the auditory nerves' or 'of or pertaining to' the noun sense of *audile*" (96).

can learn the rhythms of given x-realities, to an extent where they can make physiologically informed evaluations of a network infrastructure. That is, when a person is listening to a given network within its ambient environment, they are given information about the context (place, time, known technologies [optional], and relevant conditions), and they are left to listen to the rhythmic events and analyze a network as such. TNSs do not necessarily provide information that interferes with a person's sonic reception of a network, nor would they be accompanied by an analysis of the material. An important feature of such sonifications is their openness to public interpretation and reception, beyond the complications and structural implications of complex technological language. The hope is that with more tactical sonifications, even though sematic listening to networks would still be extremely difficult, if not impossible, people can still make sense of how technologies communicate.

Next Steps for Tactical Network Sonification

Tactical network sonification aims at the general public as its primary audience. Its interconnection to everyday life requires the technique to accommodate the various backgrounds, cultures, and interests that people may have. The main contribution of this technique is that it allows for people to listen to as well as sonify networks, without restricting one or the other to a certain group of people. For TNSs to be accessible and not intimidating—or even annoying— they would incorporate the ambient environment in their recordings. In doing so, they explicitly highlight the connection with the aspect of everyday life they are addressing, all the while informing audiences of whether or not the recorded infrastructure and environment are of interest to them. For example, a sonification that includes the ambient sounds of an office space is likely to be directed at people who work in an office setting; however, those who might be interested in the soundscape and infrastructures of office spaces might also choose to listen to and investigate such a sonification. Another essential aspect of TNSs that would reinforce their user-friendliness is a simple how-to guide for sonifying networks; however, this guide is flexible and allows people to experiment with their technique. The instruction sheet also lists the hardware and software that one must have: a microphone or recording device, an EM waves transducer, and an audio software.[39] These technical requirements are mostly affordable and readily available. Ideally, a person would be able to acquire the Detektor, or a transducer with similar capacities.

Given that the Detektor schema is open source, some may resort to making their own device. Further, the guide explains that constructing a TNS would bring together the EM acoustics, as well as the ambient sounds form the space recorded; however, people can also experiment with how they choose to sonify the infrastructure of choice. Along with the guide on how to produce TNSs, a very brief guide to listening to such sound clips helps people begin their experimenting.[40] This guide explains that rhythms and rhythmic events are at the heart of a TNS. It urges people to listen with intention, and to allow themselves to engage in an embodied experience, because this practice would allow them to access various ways in which networks may be affecting our lived everyday experience. This guide also encourages people to experiment with how they listen to a given sonification, so that their interaction with the artifact is personal and effective, as well as—ideally—affective. In focusing on one's personal experience and allowing for flexibility, the guides embrace the "tactical" in the technique, and direct the interaction to be between audiences and the artifact, as opposed to the one between audience and sonification producers.

In that respect, aesthetic features may enhance the potential of TNSs in reaching a broader public. The unpleasant nature of transduced EM waves may deter audiences from investing in listening to networks, which defeats the purpose of the technique being designed for non-experts. To address such issues, people are encouraged to experiment with making music out of network acousmatics. In other words, in addition to the environmental ambient sounds, one can mix musical notes or even songs, and reassemble the acoustic space as a musical performance. Electronic musicians for example could then include clips from tactical sonifications, and remix them into their compositions. Such a step would broaden the reach of the these sonifications to a demographic that otherwise may be difficult to reach. I have already created an account on SoundCloud—a sound publishing platform—where I will be experimenting with sharing the tactical sonifications that I produce.[41] I will also use my accounts on Twitter and Instagram, to share the soundCloud account with various circles; the two social media platforms may allow sonifications to reach various demographics. The purpose of this approach is to test how people receive the sonifications, and if and how they would interact with them. The SoundCloud currently has three examples in a TNS album.[42] It would be interesting to observe whether or not people would be attempting to create their own sonifications or experiment with the ones available.[43]

Aside from the general public, an important prospect for tactical sonifications is Environmental Studies and Earth and Ocean Sciences. In listening to the rhythms of networks,

---

[4]d with the HCMC recordings, and has used them to create a beat for a music track.

particularly when solely juxtaposed with ambient environmental sounds, these sonifications invite the potential of comparative rhythmic analyses of network soundscapes. For example, when simultaneously recording the network acousmatics of a forest and its ambient sounds, capturing bird songs, plant movement sounds, and winds, an environmental studies expert might investigate how the rhythms of the infrastructure correlate to those of the ambient environment's soundscape. While the analysis of a TNS alone may not lead to a final and determined conclusion as to the relationship between networks and environmental behaviour, the technique provides relevant information for the making of a hypothesis. Similarly, marine scientists can use the approach to investigate the effects of networks on underwater life, such as whale communications and potentially coral reefs' health. Scientists may use sonifications—of waves of various natures and frequency ranges—to determine whether tanker or recreational boating communication systems are affecting marine life, by experimenting in comparative rhythmanalysis of the various soundscapes. For example, the difference in how vibrations and waves travel above and under water may prove to be an interesting avenue for TNSs.

It is important to note that if scholars were to invest in tactical sonifications in such avenues, the technique will need to be developed to accommodate the use of high-tech, as well as incorporate mathematical techniques to arrive at precise outcomes, because in its current stages, the technique relies heavily on deep listening, as well as artistic and aesthetic methods. Although TNS is designed to share information with the general public, using listening as a tool for diagnostics is not too foreign to the scientific sphere (Sterne 2013); if they were to contribute to scholarly and academic innovation, sonifications are flexible enough to adapt scientific methods along with their rhythmanalysis focal approach.

In moving forward, TNS will continue to focus on the general public as its main audience. That said, I aim to engage with academic communities, by publishing in scholarly journals, as well as working closely with interested scholars to maintain the technique's relevance as network communications develop. Given the fast growth of network systems, I will continue to work on developing the method with experts from the technology industry, so as to account for changes that the current approach could not foresee. In this respect, I will also remain updated on scholarship in sound studies, and will attempt to develop software and hardware, as more possibilities become available. The listening technique will continue to be developed and refined, as per the findings and experiments of people—in academia or other venues. In its current form, TNS anticipates many changes, be it in hardware or practice; however, its development may rely on community engagement for sustainability and growth. The promise of tactical sonification as a listening technique for STS, specifically networks, requires a constant commitment to the study of rhythms, network communications, social and cultural effects of networks, and sound productions as anti-environments of the acousmatics of x-reality, with a consistent search for new possible implementations.

Limitations

The time and resource constraints have posed several limitations on this project. While the technique itself has been carefully designed and developed to reach the general public and provide them with a comprehensible language for addressing networks, it still requires somewhat advanced technological devices. Of all the devices one may need, the Detektor is the most difficult to acquire. It ships from Europe for free, but costs 82 Euros.[44] The schema for the

Detektor is open source, which allows for people to reproduce the device and make their own versions. This solution, however, still requires technical expertise, and might also be costly and time consuming. That said, making one's own Detektor corresponds to critical making in that people would get access to the intricacies of the design, which allows them to reflect on and possibly adjust the product as they see fit.

During the early stages of this project, and after a careful study of the Detektor's schema, I planned to remodel the device and make a transducer that would allow people to separate the networks they are detecting, as per their EM frequencies. To do so, I would have needed microchips that would capture and transduce small ranges of EM waves. After extensive research, I could not find such chips in the market, which made this approach impossible. Another option was to reverse engineer the Detektor, mapping it onto a Raspberry Pi or Arduino board, and program it so as to select the frequencies that I want to capture. The learning curve for such an exercise and the time needed to complete the prototype would have made it impossible for me to conduct the quality of research that I have presented in this project. More importantly, this limitation allowed for the deduction that studying networks holistically, prevents people from fetishizing the technical granularities of networks, embracing a cultural and sociotechnical focus on the subject matter. That said, I propose that—in the foreseeable future—a transduction device be designed and prototyped, providing affordable options for purchase. This device could provide an option to listen to networks whether as a system (the algorhythmics of x-reality), or by approaching networks in their separate capacities. The design would thus allow people to experiment with the different possibilities, reimagining tactical sonifications as network compilations in relationship to the infrastructure that houses different possibilities as a part of a whole.

In addition to the technical limitations, this dissertation does not address the issue of accessibility that a listening-based approach might impose on various communities. I acknowledge that TNS may not be accessible to everyone, as it assumes physiological hearing abilities on behalf of the listeners. I hope that as the technique evolves, it becomes more accessible to everyone. One way of becoming more inclusive requires an investment in the vibrational nature of sounds, and translating sonifications into tactile representations that people can experience. In doing so, the method would preserve its focus on and investment in embodied living, all the while allowing everyone to experience network stories within the realm of everyday life. It is important to note that a visual approach is also possible; however, this project steers its focus away from network visualisations, particularly because of the conflated notion of networks in digital studies (Venurini, Munk, and Jacomy in Vertesi and Ribes 2019). While visualization also engages the senses, sound and its affective potential—through rhythms (Duffy et al. 2011)—are vital in allowing people to experience networks as they affect our everyday life. Similarly, the politics that are inherent to visualizations are different than those of sound and listening practices, and remain outside the scope of this project. This is not to say that visual methodologies cannot or do not allow people an understanding of networks; however, this project embraces the materiality of networks, and uses sound to make tangible and sensible the physical presence of network communications. On that note, the scientific and technical details as to the physicality of networks are also outside the scope of this work, because given the context of this dissertation—within the Faculty of Humanities and aiming at the general public—relying heavily on scientific information risks replicating the problems that contemporary media often create when telling network stories.

Final Remarks

Today's everyday life has become entangled with networks and technologies that rely on a constant interaction between people and technological devices, and more importantly between technological devices themselves. The notion of x-reality (Coleman 2011) emphasizes the role of networks, and highlights the risks of pervasive media. People in general, and scholars in particular have paid close attention to the detrimental effects of network components, especially those of algorithms, and how these could reinforce social biases within our everyday life (Ananny and Crawford 2018; Crawford 2015; Kitchin 2017; Noble 2018; Sandvig et al. 2014). Addressing the infrastructures that allow for such practices to exist within our cultures and societies requires a close attention to networks, as they host these components and permit their constant and invasive operations. To access such infrastructures, people need to be equipped with proper techniques that include adequate language, while providing access to otherwise obscured operations. In response to contemporary media stories about networks, tactical network sonification is a technique that invites people to sonify and listen to communications within infrastructures, in ways that represent the rhythmic events of network operations. In doing so, this technique gives people the opportunity to actively and collectively participate in network operations, based on their material reality in our everyday life.

In conversations with people about the potential uses of TNS, the most common concern is around using the knowledge that the sonifications would provide. In accordance with tactical media's principle, and inherent to its design as an anti-environment, the role of TNS is to cause "disturbance" and invite critical thinking (Raley 6). The technique invites new ways of interacting with networks all the while placing the "emphasis on the perception of the audience" (Raley 6). In other word, in listening to a network sonification, a person is allowed freedom as to how to proceed with the knowledge with which they have just interacted. Just as with literary stories, people may enjoy sonifications for their aesthetics as much as they can learn various ways of being within the x-reality of everyday life. In a time when concerns around algorithms and networks have become a common topic of conversation, it remains essential to give people a language to discuss such issues and informedly participate in the dialogue, through methods that are as accessible and experiential as music and art.

Works Cited

"0.1 GHz to 2.5 GHz 70 dB Logarithmic Detector/Controller – Data Sheet." *Analog Devices*, 1998-2015, www.analog.com/media/en/technical-documentation/data-sheets/AD8313.pdf.

"Advanced Research Projects Agency Network (ARPANET)." *Techopedia*, 2 Feb. 2017, www.techopedia.com/definition/2381/advanced-research-projects-agency-network-arpanet.

Agre, Philip E. *Computation and Human Experience*. Cambridge U P, 1997.

Aneesh, A. *Virtual Migration: The Programming of Globalization*. Duke U P, 2006.

Ananny, Mike, and Kate Crawford. "Seeing without Knowing: Limitations of the Transparency Ideal and its Application to Algorithmic Accountability." *New Media & Society*, vol. 20, no. 3, 2018, pp. 973-89.

"ARPANET." *ComputerHope*, 16 Jun. 2017, www.computerhope.com/jargon/a/arpanet.htm.

Audacity (Software). Audacity®, 2020. https://www.audacityteam.org/.

Balsillie, Jim. "Sidewalk Toronto has only one beneficiary, and it is not Toronto." *The Globe and Mail*, 29 Nov. 2018, www.theglobeandmail.com/opinion/article-sidewalk-toronto-is-not-a-smart-city/.

Barthes, Roland. "Listening." In *The Responsibility of Forms: Critical Essays on Music, Art, and Representation*, Translated by Richard Howard, U of California P, 1991, 245-60.

Buechley, Leah and Michael Eisenberg. "The LilyPad Arduino: Toward Wearable Engineering for Everyone." *Pervasive Computing*, edited by Jadwiga Indulska, Donald Patterson, Tom Rodden, and Max Ott, Springer-Verlag, 2008, pp. 12-15.

Burke, Dennis. "All Circuits are Busy Now: The 1990 AT&T Long Distance Network Collapse."

    CSC440-01. California Polytechnic State University, 1995,

    https://users.csc.calpoly.edu/~jdalbey/SWE/Papers/att_collapse.

Cannam, Chris, Christian Landone, and Mark Sandler. "Sonic Visualiser: An Open Source

    Application for Viewing, Analysing, and Annotating Music Audio Files." In *Proceedings*

    of the ACM Multimedia 2010 International Conference, 2010.

Canon, Gabrielle. "'City of surveillance': Privacy Expert Quits Toronto's Smart-City Project."

    The Guardian, 23 Oct. 2018, www.theguardian.com/world/2018/oct/23/toronto-smart-

    city-surveillance-ann-cavoukian-resigns-privacy.

Chion, Michel. *Audio-Vision: Sound on Screen*. Columbia U P, 1994.

Christensson, Per. "Torrent Definition." *TechTerms*. Sharpened Productions, 03 Oct.2007,

    https://techterms.com/definition/torrent.

Chun, Wendy Hui Kyong. "Crisis, Crisis, Crisis, or Sovereignty and Networks." *Theory,*

    Culture, & Society, vol. 28, no. 6, 2011, pp. 91-112.

Clarke, Arthur C. *Profiles of the Future: An Inquiry into the Limits of the Possible*. Harper and

    Row, 1973.

Coleman, Beth. *Hello Avatar*. MIT P, 2011.

---. "Smart Things, Smart Subjects: How the 'Internet of Things' Enacts Pervasive Media." In

    The Routledge Computation to Media Studies and Digital Humanities, edited by Jentery

    Sayers, 222-229. Routledge, 2018.

Coletta, Claudio, and Rob Kitchin. "Algorhythmic governance: Regulating the 'heartbeat' of a

    city using the Internet of Things." *Big Data & Society*, Dec. 2017,

    doi:10.1177/2053951717742418.

Crawford, Kate. "Can an Algorithm Be Agonistic? Ten Scenes from Life in Calculated Publics."

    Science, Technology & Human Values, vol. 41, no. 1, 2015, pp. 77-92.

Dean, Jodi. "Communicative Capitalism: Circulation and the Foreclosure of Politics." *Cultural*

    Politics, vol. 1, no. 1, 2005, pp: 51-74.

Dieter, Michael. "The Virtues of Critical Technical Practice." *Differences*, vol. 25, no. 1, 2014,

    pp. 216-30.

DiSalvo, Carl. *Adversarial Design*. MIT P, 2012.

Doctoroff, Dan. "Sidewalk Toronto Project Update." *Medium-SideWalkTalk*. 14 Feb. 2019.

    https://medium.com/sidewalk-toronto/sidewalk-toronto-project-update-d44738cdb239

---. "Why We're No Longer Pursuing the Quayside Project—and What's Next for Sidewalk

    Labs." *Medium-SideWalkTalk*. 7 May 2020, www.medium.com/sidewalk-talk/why-were-

    no-longer-pursuing-the-quayside-project-and-what-s-next-for-sidewalk-labs-

    9a61de3fee3a.

Duffy, Michelle, Gordon Waitt, Andrew Gorman-Murray, and Chris Gibson. "Bodily Rhythms:

    Corporeal Capacities to Engage with Festival Places." *Emotion, Space, and Society* vol.

    4, 2011, pp. 17-24.

Edensor, Tim. "The Sonic Rhythms of Place." *The Routledge Companion to Sound Studies*,

    edited by Michael Bull. Routledge, 2018, pp. 158-65.

Friedman, Batya and Peter H. Kahn, Jr. "Human Agency and Responsible Computing:

    Implications for Computer System Design." In *Human Values and the Design of*

    Computer Technology, edited by Batya Friedman. CSLI Publications, 1997, pp. 221-35.

Galloway, Alexander R. *The Interface Effect*. Polity Press, 2012.

---. *Protocols: How Control Exists After Decentralization*. MIT P, 2004.

GarageBand (Software). Apple Inc, 2020. https://www.apple.com/ca/mac/garageband/.

Gawron, Nicole. "Infamous Software Bugs: AT&T Switches," *Olenick and Associates, Inc.* 5

Sept. 2014, https://www.olenick.com/blog/articles/infamous-software-bugs-at-t-

switches/.

Gillespie, Tarleton. "Algorithm." *Digital Keywords: A Vocabulary of Information Society and*

Culture, edited by Benjamin Peter. Princeton U P, 2016, pp. 18-30.

Gow, Gordon, and Richard Smith. *EBOOK: Mobile and Wireless Communications: An*

Introduction. McGraw-Hill Education, 2006. ProQuest Ebook Central,

http://ebookcentral.proquest.com/lib/uvic/detail.action?docID=290393.

"The Guardian view on Google and Toronto: smart city, dumb deal (Editorial)." *The Guardian*, 5

Feb. 2018, https://www.theguardian.com/commentisfree/2018/feb/05/the-guardian-view-

on-google-and-toronto-smart-city-dumb-deal.

Halt and Catch Fire. Created and produced by Christopher Cantwell and Christopher C. Rogers.

Netflix, 2017.

Jackson, Steven J. "Rethinking Repair." In *Media Technologies*, edited by Tarleton Gillespie,

Pablo J. Boczkowski, and Kirsten A. Foot. MIT P, 2014, pp. 221-40.

Jay, Martin. *Downcast Eyes: The Denigration of Vision in Twentieth-Century French Thought*.

UC P, 1993.

Kirschenbaum, Matthew G. *Mechanisms: New Media and the Forensic Imagination*. MIT P,

2007.

Kitchin, Rob. "Thinking critically about and researching algorithms." *Information,*

Communication, and Society, vol. 20, no.1, 2017, pp. 14–29.

http://dx.doi.org/10.1080/1369118X.2016.1154087

Kleinrock, Leonard. *Communication Nets: Stochastic Message Flow and Delay*. McGraw-Hill

Book Company, 1964.

Kotecki, Peter. "Google's Parent Company Revealed its Plan for a High-Tech Neighborhood in

Toronto – and it Could Be the World's Largest Tall Timber Project." *Business Insider*, 16

Aug. 2018, www.businessinsider.in/googles-parent-company-revealed-its-plan-for-a-

high-tech-neighborhood-in-toronto-and-it-could-be-the-worlds-largest-tall-timber-

project/articleshow/65417384.cms.

Krapp, Peter. *Noise Channels: Glitches and Error in Digital Culture*. University of Minnesota P,

2011.

Latour, Bruno. *Science in Action: How to Follow Scientists and Engineers Through Society*.

Harvard U P, 1988.

---. "Why Has Critique Run out of Steam? From Matters of Fact to Matters of Concern." *Critical

Inquiry*, vol. 30, Winter 2004, pp. 225- 48.

Lefebvre, Henri. *The Production of Space*. Translated by Donald Nicholson-Smith. Blackwell

Publishers, 1991.

---. *Rhythmanalysis: Space, Time and Everyday Life*. Translated by Stuart Elden and Gerald

Moore. Bloomsbury Revelations, 2013.

Lessig, Lawrence. *Code and Other Laws of Cyberspace*. Basic Books, 2000.

Lyon, Dawn. *What is Rhythmanalysis*. Bloomsbury Academic, 2019.

Mackenzie, Adrian. *Wirelessness: Radical Empiricism in Network Cultures*. MIT P, 2010.

Marino, Mark C. "Reading Culture through Code." In *The Routledge Companion to Media

Studies and Digital Humanities*, edited by Jentery Sayers. Routledge, 2018, pp. 472-82.

Marvin, Carolyn. *When Old Technologies Were New: Thinking about Electric Communication in the Late Nineteenth Century*. Oxford U P, 1988.

Mattern, Shannon. "Sonic Archaeologies." In *The Routledge Companion to Sound Studies*, edited by Michael Bull. Routledge, pp. 222-30, 2018.

---. "Urban Auscultation: or, Perceiving the Action of the Heart." *Places Journal*, April 2020, https://placesjournal.org/article/urban-auscultation-or-perceiving-the-action-of-the-heart/.

McGann, Jerome. *The Romantic Ideology: A Critical Investigation*. University of Chicago P, 1983.

McLuhan, Marshall, and Quentin Fiore. *The Medium Is the Massage*. Random House of Canada Limited, 1996.

McPherson, Tara. "Introduction: Media Studies and the Digital Humanities." *Cinema Journal*, vol. 48, no. 2, 2009, pp. 119-23.

Mecava, Aridan. "Sidewalk Toronto: Tech-Utopia, Authentic Neighborhood, Or Disaster?" *Popup City*, 7 Nov. 2017, https://popupcity.net/observations/sidewalk-toronto-tech-utopia-authentic-neighborhood-or-disaster/.

Miyazaki, Shintaro. "Algorhythmics: A diffractive Approach for Understanding Computation." *The Routledge Computation to Media Studies and Digital Humanities*, edited by Jentery Sayers. Routledge, 2018, pp. 243-49.

---. "Algorhythmics: Understanding Micro-Temporality in Computational Cultures." *Computational Culture* vol. 2, 28 Sep. 2012, http://computationalculture.net/algorhythmics-understanding-micro-temporality-in-computational-cultures/.

---. "Algorhythmic Ecosystems: Neoliberal Couplings and Their Pathologies 1960-Present."

Algorithmic Cultures: Essays on Meaning, Performance and New Technologies, edited

by Robert Seyfert and Jonathan Roberge. Routledge, 2016, pp. 128-139.

---. "Going Beyond the Visible: New Aesthetic as an Aesthetic of Blindness?" In *Postdigital*

Aesthetics, edited by David. M. Berry and Michael Dieter. Palgrave Macmillan, pp. 219-

31, 2015.

---. "Urban Sounds Unheard-of: A Media Archaeology of Ubiquitous Infospheres." *Continuum:*

Journal of Media & Cultural Studies, vol. 27, no. 4, 2013, pp. 514-22.

Mouffe, Chantal. *The Democratic Paradox*. Verso, 2000.

Mullen, Don. "AT&T Pinpoints Cause of Long Distance Line Crash." *UPI*, 17 Jan. 1990,

www.upi.com/Archives/1990/01/17/ATT-pinpoints-cause-of-long-distance-line-

crash/7025632552400/.

Muoio, Danielle. "Google wants to build its own futuristic, smart city." *Business Insider*, 27 Apr.

2016, https://www.businessinsider.com/google-wants-to-build-its-own-smart-city-2016-

4.

The National Computing. *Introducing Communications Protocols*. NCC Publications, 1978.

Noble, Safiya Umoja. *Algorithms of Oppression: How Search Engines Reinforce Racism*. New

York U P, 2018.

Oliveros, Pauline. *Deep Listening: A Composer's Sound Practice*. Lincoln, NE: iUniverse, 2005.

---. "Some Sound Observations." *Audio Culture: Readings in Modern Music*, edited by Cox,

Christoph and Daniel Warner. Continuum Publishing, 2010, pp. 102-6.

Palmer, Michaela, and Owain Jones. "On Breathing and Geography: Explorations of Data

Sonifications of Timespace Processes with Illustrating Examples from a Tidally Dynamic

Landscape (Severn Estuary, UK)." *Environment and Planning*, vol. 46, 2014, pp. 222-40.

Penny, Simon. "Agents as Artworks and Agent Design as Artistic Practice." In *Human Cognition

and Social Agent Technology* edited by Kerstin Dautenhahn. Amsterdam: John

Benjamins, 1999. http://simonpenny.net/1990Writings/agentdesign.html.

"Plan Development Agreement Between Toronto Revitalization Corporation and Sidewalk Labs

llc." *Quayside*, July 2018. https://quaysideto.ca/wp-content/uploads/2020/05/Plan-

Development-Agreement-Amendments-and-Sidewalk-Labs-Termination-Notice-

Termination-Effective-May-17-2020.pdf.

Pringle, Ramona. "Google Plans to 'Fix' Toronto by Building Smart City." *CBC*, 20 May 2017,

www.cbc.ca/news/technology/google-smart-city-toronto-1.4123289.

Quayside Civic Labs. "Data Collection and Use." *The Quayside Civic Labs Info Sheet Series*, 1-

4, n.d., https://waterfrontoronto.ca/nbe/wcm/connect/waterfront/f03805a6-6f08-489b-

9e9e-e1bf6f06d3eb/Civic+Labs+-+Quayside+-

+Info+Sheet+3+(Data+Collection+and+Use)+-

+December+12%2C+2018.pdf?MOD=AJPERES&CONVERT_TO=url&CACHEID=f03

805a6-6f08-489b-9e9e-e1bf6f06d3eb.

Rajko Jessica. "A Call to Action: Embodied Thinking and Human-Computer Interaction

Design." In *The Routledge Companion to Media Studies and Digital Humanities*, edited

by Jentery Sayers, 195-203. London: Routledge, 2018.

Raley, Rita. *Tactical Media*. U of Minnesota P, 2009.

Ratto, Matt. "Critical Making: Conceptual and Material studies in Technology and Social Life."
The Information Society, vol. 27, no. 3, 2011, pp. 252-60.

Rodgers, Tara. "Approaching Sound." In *The Routledge Computation to Media Studies and Digital Humanities*, edited by Jentery Sayers, 233-242. Routledge, 2018.

Rosner, Daniela K. and Morgan G. Ames. "Designing for Repair? Infrastructures and Materialities of Breakdown," *CSCW'14 Proceeding of the 17th ACM Conference on Computer Supported Cooperative Work and Social Computing*, 2014, pp. 319-331.

Rouse, Margaret. "ARPANET." *SearchNetworking*. Mar. 2017,
www.searchnetworking.techtarget.com/definition/ARPANET.

Sandvig, Christian, Kevin Hamilton, Karrie Karahalios, and Cedric Langbort. "Auditing Algorithms: Research Methods for Detecting Discrimination on Internet Platforms." *Computational Culture*, 2014, http://www-personal.umich.edu/~csandvig/research/Auditing%20Algorithms%20--%20Sandvig%20--%20ICA%202014%20Data%20and%20Discrimination%20Preconference.pdf.

Schafer, R. Murray. *The Soundscape: Our Sonic Environment and the Tuning of the World*. Destiny Books, 1994.

Sengers, Phoebe. "The Engineering of Experience." In *Funology*, edited by Mark A. Blythe, Kees Overbeeke, Andrew F. Monk, and Peter C. Wright. Springer, 2005, pp. 19-29.

Sengers, Phoebe, Kirsten Boehner, Shay David, and Joseph 'Jofish' Kaye. "Reflective Design." *Proceedings of the 4th Decennial Conference on Critical Computing between Sense and Sensibility—CC '05*, edited by Olav W. Bertelsen, Niels Olof Bouvin, Peter G. Krogh, and Morten Kyng. ACM P, 2005, pp. 58-59.

Sharma, Sarah. "Many McLuhan or None at All." *Canadian Journal of Communication*, vol. 44,

> no. 4, 2019, pp. 483-488.

Shaviro, Steven. *No Speed Limit: Three Essays on Accelerationism*. University of Minnesota P,

> 2015, https://manifold.umn.edu/read/54023bc6-0a45-4a51-9fc3-
> 186e9e512a61/section/1b09b41b-002e-4b2d-ade0-b7ee76e5e9cc.

Sidewalk Labs. 2020. Sidewalklabs.com

Skiesbleed. *SoundCloud*. https://soundcloud.com/skiesbleed.

The Social Dilemma. Directed by Jeff Orlowski. Exposure Labs, 2020.

Sonic Archaeology. "0016 HF-Backgroundnoise." *SoundCloud*. 2010,

> https://soundcloud.com/sonic-archeology/0016-hf-backgroundnoise.

Sonic Archaeology. "0018 HF-pc lookingforwifi." *SoundCloud*. 2010,

> https://soundcloud.com/sonic-archeology/0018-hf-pc-lookingforwifi.

Sonic Archaeology. "GSM-Incoming-call." *SoundCloud*. 2010, https://soundcloud.com/sonic-

> archeology/gsm-incoming-call.

Sonic Archaeology. "Torrent-transmission." *SoundCloud*. 2012, https://soundcloud.com/sonic-

> archeology/torrent-transmission.

Sonic Visualizer (Software). Centre for Digital Music. Queen Mary, University of London, 2020.

> https://www.sonicvisualiser.org/.

Sterne, Jonathan. *The Audible Past: Cultural Origins of Sound Reproduction*. Duke U P, 2003.

---, editor. *The Sound Studies Reader*. Routledge, 2012.

Strubell, Emma, Ananya Ganesh, Andrew McCallum. "Energy and Policy Considerations for

> Deep Learning in NLP." *The 57th Annual Meeting of the Association for Computational

> Linguistics (ACL)*, 2019, https://arxiv.org/pdf/1906.02243.pdf.

Thurm, Eric. "Farewell to Halt and Catch Fire, the Best Show that Nobody Watched." *The Guardian*, 16 Oct. 2017, https://www.theguardian.com/tv-and-radio/2017/oct/16/farewell-to-halt-and-catch-fire-the-best-show-that-nobody-watched.

Tumulty, Karen. "AT&T Reaches Out to Public and Apologizes for Breakdown: Telecommunications: The Firm Suspects that a 'Bug' in Computer Software Caused its System's Collapse. It May Offer Restitution to Some and a Day of Discounted Rates to All." *Los Angeles Times*, 17 Jan. 1990, www.latimes.com/archives/la-xpm-1990-01-17-fi-215-story.html.

Vee, Annette. "Programming Literacy." *The Routledge Companion to Media Studies and Digital Humanities*, edited by Jentery Sayers. Routledge, 2018, pp. 445-52.

Vertesi, Janet and David Ribes, editors. *digitalSTS: A Field Guide for Science and Technology Studies*. Princeton U P, 2019.

@WaterfrontTO. "The Quayside project has prompted many questions around data collection, storage, use and consent. Check out this Info Sheet where we address some of these questions: http://ow.ly/JmgO30nSZAe." *Twitter*, 03 Mar. 2019. https://twitter.com/WaterfrontTO/status/1102323352549548032.

Weiser, Mark. "Ubiquitous Computing." *Hot Topics* edited by Ronald D. Williams: Oct. 1993, pp. 71-2.

Weiser, Mark and John Seely Brown. "The Coming Age of Calm Technology." In *Beyond Calculation:* 75-85. Springer, 1996.

Wernimont, Jacqueline. "The Living Net." *Jwernimont*. 7 May 2018, https://jwernimont.com/the-living-net/.

Zimmermann, Kim Ann, and Jesse Emspak. "Internet History Timeline: ARPANET to the

World Wide Web." *LiveScience*. 27 Jun. 2017, https://livescience.com/20727-internet-

history.html.

Appendices

Appendix A: A Guide for Tactical Network Sonification

This guide explains how to produce a tactical network sonification (TNS), in a way that maintains people's autonomy, without dictating one specific path of production over the other. It does, however, provide a list of devices and software necessary for the producing of a TNS. The general idea of a TNS is to produce a sonification of an infrastructure. People may also want to mix together EM acoustics and ambient natural and environmental sounds. The particulars of such sonifications and their aesthetics are completely left for people to experiment with and produce what best represents the system with which they are working. The important notion to keep in mind is that the main purpose of a tactical sonification is to make networks apprehendable to people; anything beyond that is up to the sonification producer and their audience's reception.

Hardware

To record the sounds needed for a TNS, people need a microphone and a Detektor. The microphone would be used to record ambient sounds, and the Detektor to transduce and record EM waves. To have good quality recordings, people are encouraged to use an audio interface, to regulate the input sounds and levels captured. The Detektor schema is openly available online, for those who would like to make their own transducer devices.[45] In addition to the recording kit, people will need a computer, and possibly some music instruments, if music is the direction they will be taking.

Software

To record and listen to clips, TNS producers will need a sound software. The choice of software is dependent of the person's experience with such tools, and what they are wanting to do. For the purposes of this dissertation, I have used Audacity, Sonic Visualiser, and Apple's GarageBand.[46]

Appendix B: A Guide for Listening to Tactical Network Sonification

Listening to a tactical network sonification (TNS) entails attention to the rhythms of the clip in question, with the intention of understanding it beyond the mere sounds to which one is listening. Further, to benefit the most from a TNS, listeners are encouraged to embrace the embodied experience that will occur during a listening session. In other words, if the sound clip makes one's body tense up, then it is worth questioning to what effect such experiences are mimicking what we might be experiencing unknowingly, and whether or not other living organisms experience such effects directly from the EM waves that the recorded network infrastructure produces. When listening to a TNS, listeners are also encouraged to experiment with their approach; for instance, they can try to listen to the same clip in different settings, or listen to it once with headphones and once with speakers. Another method would be changing the speed of the sound clip, to examine how such changes would affect how one's physical response changes. People may also choose to listen to a clip backwards, which might produce some interesting findings. The listener's experience is at the heart of TNS's principles and values, whereby producers of these sonifications design their clips with a flexibility granted to the listener as they experiment with and experience the various sounds, in the different methods they choose to adopt.

9 783384 222138